Symmetry and Liquid Crystals

Symmetry and Liquid Crystals

Editors

Shoichi Ishihara
Sadahito Uto

MDPI • Basel • Beijing • Wuhan • Barcelona • Belgrade • Manchester • Tokyo • Cluj • Tianjin

Editors
Shoichi Ishihara
Department of Electrical and Electronic Engineering
Osaka Institute of Technology
Osaka
Japan

Sadahito Uto
Department of Biomedical Engineering
Osaka Institute of Technology
Osaka
Japan

Editorial Office
MDPI
St. Alban-Anlage 66
4052 Basel, Switzerland

This is a reprint of articles from the Special Issue published online in the open access journal *Symmetry* (ISSN 2073-8994) (available at: www.mdpi.com/journal/symmetry/special_issues/Symmetry_Liquid_Crystals_LC).

For citation purposes, cite each article independently as indicated on the article page online and as indicated below:

LastName, A.A.; LastName, B.B.; LastName, C.C. Article Title. *Journal Name* **Year**, *Volume Number*, Page Range.

ISBN 978-3-0365-8323-5 (Hbk)
ISBN 978-3-0365-8322-8 (PDF)

Cover image courtesy of Shoichi Ishihara

Contents

About the Editors

Shoichi Ishihara

Shoichi Ishihara, PhD in engineering, was a professor at Osaka Institute of Technology (OIT) from 2010 to 2021. He is a former president and honorary member of the Japanese Liquid Crystal Society (JLCS) and a member of the Society for Information Display (SID). He obtained his BS in synthetic chemistry at Nagoya Institute of Technology in 1973 and his PhD in physics and electronics at Osaka Prefecture University in 2000. He joined Matsushita Electric Industrial Co, Ltd, currently Panasonic Corporation, in 1973. From 1978 to 1979, he was a fellow at Massachusetts Institute of Technology, Boston, USA. From 1996 to 2002, his activities have focused on the alignment control of LCs and the optical design of retardation films. There, he had developed optically compensated birefringence (OCB)-mode LCDs and put them into the practical use for the first time. From 2003 to 2010, he worked at Sharp Corporation where he was the first to put photoaligned LCDs and transverse bend alignment (TBA)-mode LCDs to practical use in the world. His special field is display mode and interaction between LC molecules and alignment layers. The optical design of LCD is also his strong area.

At OIT, he led a novel gas sensing project using cholesteric LCs with a consumer electronics maker, and also led a developing project of a film substrate for flexible LCDs. He holds over 130 Japanese patents and over 42 US patents and has published over 50 original papers.

The recently published book "High Quality Liquid Crystal Displays and Smart Devices", edited by S. Ishihara et al. (the Institution of Engineering and Technology (England), 2019), and the paper "Alignment control technology of liquid crystal molecules" (J. Soc. Inf. Display, Vol.28, pp.44–74, 2020) won the JLCS Excellent Book Award and the JLCS Best Paper Award, respectively.

Sadahito Uto

Since 2017, Sadahito Uto has been working as a professor at the Department of Biomedical Engineering, the Osaka Institute of Technology. He graduated from the Faculty of Engineering, Ehime University, and received his master's degree in 1994 on the study on the Kerr effect in liquid dielectrics. After that, he went on to graduate at Osaka University and got his Ph.D. He prepared free-standing films of ferroelectric liquid crystals and the first non-solid ferroelectrics, and studied their electrical and optical properties. He is also working on the development of tunable color devices and ion sensor devices using the selective reflection of hydroxypropyl cellulose liquid crystals.

Besides liquid crystals, he also studied the electrical properties of nerves and muscles. For half a year in 2001, he was a visiting researcher at Harvard University in the United States. He has proposed a method to identify grayanotoxin derivatives, a class of neurotoxins, using the applied electric-field-induced contraction of cultured muscle derived from mouse skeletal muscle.

Preface to "Symmetry and Liquid Crystals"

Liquid crystals (LCs) are aggregates of individual molecules due to moderate intermolecular interactions, and are roughly divided into lyotropic LCs and thermotropic LCs. The latter includes smectic, nematic, cholesteric, and discotic phases, depending on the molecular arrangement and symmetry, and changes within the external field such as temperature and pressure.

Currently, the scale of the global LC industry is approximately USD 115 billion, most of which is for display applications using nematic LCs (including chiral nematic LCs). The research and development of cholesteric and smectic LCs are also actively carried out for sensors, organic semiconductors, solar cells, and so on.

Generally, a LC device confines LC molecules in a cell and achieves a desired LC alignment via interfacial force. LCs are inherently anisotropic media, causing symmetry/asymmetry configuration, and can be controlled by an external field and used as an optical or electrical device. The purpose of this Special Issue is to provide new knowledge and research tools as the application of LC devices expands from flat displays to new fields such as 3D displays and sensors.

In this Special Issue entitled "Symmetry and Liquid Crystals", 11 articles are collected. There, we will focus on the symmetry of the chemical structure of LC molecules, the asymmetry of the alignment process at the substrate interface, and the asymmetry of the electric field applied to the LC layer. This Special Issue will provide LC researchers with ways to solve their problems.

This Special Issue contains five articles that focus on the chemical structure of LC materials and dopants, where relating polymer LCs and elastomers are also mentioned. There, the relationship between chemical structure and physical properties has been experimentally and theoretically investigated in detail, and applications to light emitters, solar cells, actuators, etc., have been proposed.

With the increase in the size and definition of liquid crystal displays (LCDs) and the expansion of applications such as smart glasses and three-dimensional displays, the demand for high-speed response and low-voltage driving is increasing. In this Special Issue, we can find another four articles on LC devices which utilize blue-phase LC materials or interfacial alignment control technology to meet these needs. Also, in one article of this issue, a unique approach to achieving high-speed response by combining a normal nematic LCD with a dyanamic retarder is discussed.

In addition, in the in-plane switching (IPS) mode and fringe-field switching (FFS) mode, it was difficult to evaluate the residual DC voltage (VrDC), which is a quantitative evaluation index for image sticking generated by various factors. In this Special Issue, a new evaluation method for this VrDC is proposed.

We want to express our great appreciation for the efficient assistance provided by Amelia Sun and the Editorial Office of *Symmetry*.

Shoichi Ishihara and Sadahito Uto
Editors

symmetry

MDPI

view

hiral Liquid Crystalline Electronic Systems

ısahiro Funahashi [1,2]

[1] Program for Advanced Materials Science, Faculty of Engineering and Design, Kagawa University, 2217-20 Hayashi-cho, Takamatsu, Kagawa 761-0396, Japan; funahashi.masahiro@kagawa-u.ac.jp; Tel.: +81-87-864-2411

[2] Health Research Institute, National Institute of Advanced Industrial Science and Technology, 2217-14 Hayashi-cho, Takamatsu, Kagawa 761-0395, Japan

Abstract: Liquid crystals bearing extended π-conjugated units function as organic semiconductors and liquid crystalline semiconductors have been studied for their applications in light-emitting diodes, field-effect transistors, and solar cells. However, studies on electronic functionalities in chiral liquid crystal phases have been limited so far. Electronic charge carrier transport has been confirmed in chiral nematic and chiral smectic C phases. In the chiral nematic phase, consisting of molecules bearing extended π-conjugated units, circularly polarized photoluminescence has been observed within the wavelength range of reflection band. Recently, circularly polarized electroluminescence has been confirmed from devices based on active layers of chiral conjugated polymers with twisted structures induced by the molecular chirality. The chiral smectic C phase of oligothiophene derivatives is ferroelectric and indicates a bulk photovoltaic effect, which is driven by spontaneous polarization. This bulk photovoltaic effect has also been observed in achiral polar liquid crystal phases in which extended π-conjugated units are properly assembled. In this manuscript, optical and electronic functions of these chiral π-conjugated liquid crystalline semiconductors are reviewed.

Keywords: liquid crystal; liquid crystalline semiconductor; chiral nematic phase; chiral smectic C phase; ferroelectric liquid crystals; circularly polarized light; circularly polarized photoluminescence; circularly polarized electroluminescence; bulk photovoltaic effect; shift current

ation: Funahashi, M. Chiral
ɩuid Crystalline Electronic Systems.
ımetry **2021**, *13*, 672. https://
org/10.3390/sym13040672

ademic Editors: Shoichi Ishihara
: Sadahito Uto

eived: 22 March 2021
epted: 10 April 2021
olished: 13 April 2021

1. Introduction

1.1. Chiral Liquid Crystalline Phases

Molecular chirality often breaks the symmetry of liquid crystalline (LC) phases to induce twisted structures or electrical polarization. In the chiral nematic (N*) phase, the nematic order is twisted along an axis perpendicular to the director, as shown in Figure 1a [1]. The helical structure of the N* phase interacts with circularly polarized (CP) light. When the wavelength of the incident light coincides with the helical pitch of the N* phase, the CP light with the same handedness of the N* phase is reflected and that with the opposite handedness is transmitted through the N* phase (Figure 1a). The wavelength reflected from the N* phase, λ is described in Equation (1), where p, n and θ are helical pitch, refractive index, and incident angle, respectively.

$$2\left(\frac{p}{2}\right)n\sin\theta = m\lambda \quad (1)$$

The width of reflection band $\Delta\lambda$ is determined by the helical pitch and birefringence Δn, as expressed in Equation (2).

$$\Delta\lambda = p\cdot\Delta n \quad (2)$$

If fluorescent chromophores are doped in an N* film with a left-handed helical structure, a right-handed CP fluorescence is emitted from the sample and a left-handed one is confined in the helical structure within the wavelength range of the reflection band of the N* phase, as shown in Figure 1b.

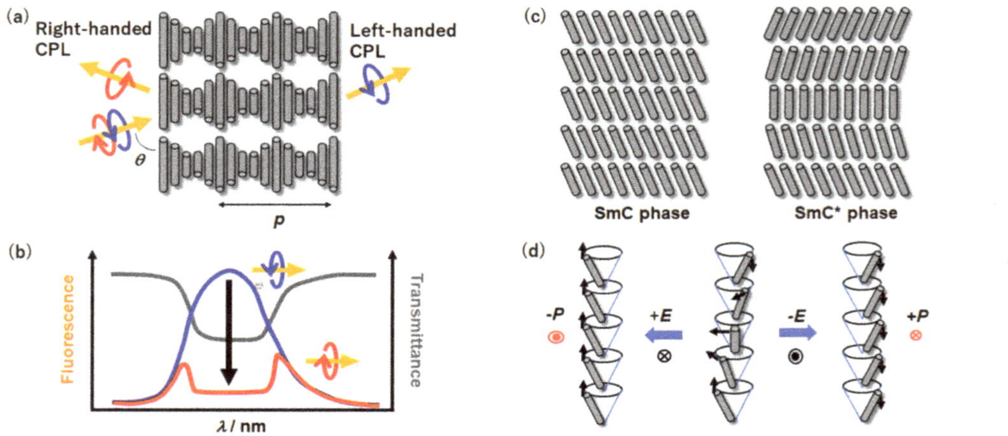

Figure 1. (**a**) Schematic illustration for supramolecular structures of a N* phase and selective reflection. Red and blue arrows denote right- and left-handed CP light, respectively. (**b**) Relationship between a reflection band and CP PL spectra. (**c**) Schematic illustration for molecular aggregation structures of SmC and SmC* phases. Red and blue lines indicate right- and left-handed handed CP fluorescence spectra, respectively. Gray line denotes transmittance spectrum. (**d**) Schematic illustration for spontaneous polarization in an SmC* phase.

In the smectic phase in which the director tilts from the layer normal, molecu chirality induces a twisted structure, as shown in Figure 1c. In a smectic C (SmC) pha consisting of achiral molecules, the director tilts from the layer normal and no positio order within a layer. In the chiral smectic C (SmC*) phase formed by chiral LC molecu the director changes continuously along the layer normal so that the director draws helix [2]. In the twisted state, an electrical polarization is canceled. As shown in Figure a DC voltage is applied in the direction perpendicular to the layer normal, the twist structure is unwounded to induce macroscopic electrical polarization, which does r disappear after the removal of the DC bias. Therefore, the SmC* phase is ferroelect Macroscopic electrical polarization can also be induced in the more ordered smectic pha by the tilted director from the layer normal and chirality.

Blue phases [3] and twisted-grain-boundary (TGB) phases [4] appear under strong twisted conditions and formation of defects plays a significant role. Various unique opti and electro-optical phenomena have been studied, although these phases are electrica insulative and the author does not mention these phases in this review.

1.2. Liquid Crystalline Semiconductors

While conventional liquid crystals are electrical insulators, liquid crystals comprisi extended π-conjugated units exhibit electronic charge carrier transport, which is observ in semiconductors [5–8]. Figure 2 shows molecular structures of typical LC semiconducto Electronic charge carrier transport has been confirmed in columnar and calamitic LC s tems. Electroactive π-conjugated moieties are self-organized in columnar or layer structu to form one or two-dimensional electronic systems. The electronic carrier transport the smectic or columnar phases proceeds by a charge hopping mechanism consideri energetic and positional disorders [9], although bandlike behavior was observed in dered smectic and ordered columnar phases [10]. From the viewpoint of supramolecu chemistry, LC assemblies can be regarded as a field for integration of electronic functio of π-conjugated units [11].

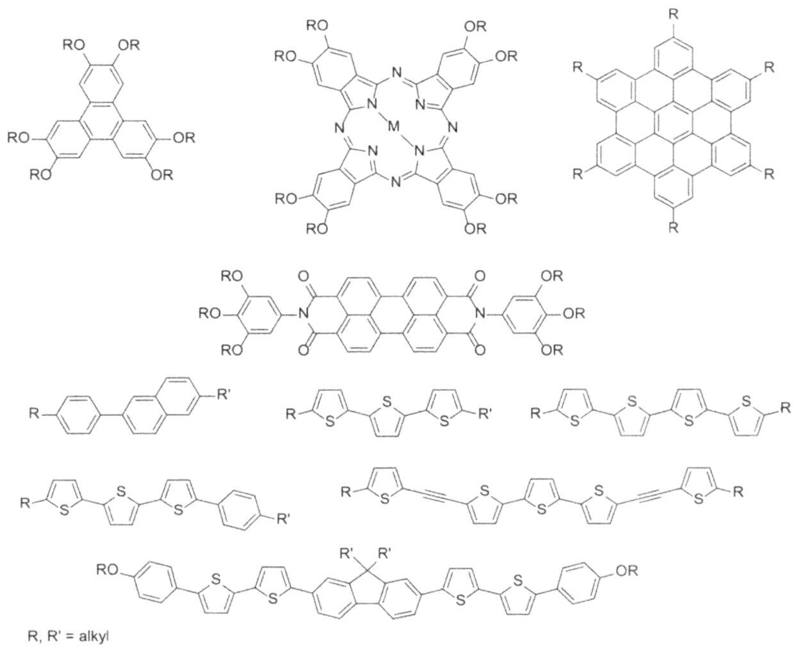

Figure 2. Molecular structures of typical liquid crystalline semiconductors.

As shown in Figure 2, triphenylene, hexabenzocoronene, phthalocyanine, perylene bisimide derivatives are known as columnar LC semiconductors [12–16]. Phenylnaphthalene and oligothiophene derivatives exhibit good carrier transport properties in smectic phases [17–21]. The application of LC semiconductors to electroluminescence device [6,22], field effect transistors [5,23–27], and solar cells have been investigated [28–30].

In the LC phases with a nematic order, ionic conduction had been predominant. However, electronic charge carrier transport has been confirmed in the nematic phases consisting of the rod-like and disk-like molecules comprising extended π-conjugated cores [31–34]. In chiral LC phases, electronic charge carrier transport should be possible if π-conjugated moieties aggregate closely, resulting in a large overlap between the π-conjugated units.

1.3. Chiral Liquid Crystalline Electronic Systems

Thus far, most of the studies on LC semiconductors have been limited to achiral systems. In liquid crystal phases, molecular chirality can induce supramolecular periodicity or electrically polarized structures, as mentioned in Section 1.1 Chiral liquid crystalline electronic systems can be constructed by assembling electroactive π-conjugated units in the chiral LC phase. The electronic function of the assemblies of the π-conjugated units is coupled with the supramolecular structures induced by molecular chirality to produce new optoelectronic properties. In this article, CP light emission in the chiral nematic phase and bulk photovoltaic effect in the ferroelectric phases of LC molecules comprising extended π-conjugated units are reviewed. Efficient CP light emitters are required for display and telecommunication applications. The bulk photovoltaic effect is driven by the internal electric field of ferroelectrics and has a potential to generate higher open-circuit voltages than the band gaps. These topics are related to new applications of electroactive liquid crystals beyond conventional display applications.

2. Electronic Systems with the Chiral Nematic Order

The N* phase indicates anisotropic transmission for CP light and can be used for C light emitters. Conventional N* liquid crystals are electrical insulators. CP light emissic and lasing by optical pumping methods in dye-doped cholesteric liquid crystals have bee studied using thick samples in the order of 10 to 100 μm [35–38]. For the constructic of electrical-pumping systems based on N* liquid crystals, high density of π-conjugate chromophores is required for efficient electronic charge carrier transport. In this section, C light emissions from N* liquid crystals containing extended π-conjugated chromophor are reviewed.

2.1. Circularly Polarized Photoluminescence from N Liquid Crystals*

Chen et al. reported high quality CP photoluminescence (PL) from dye-doped N LCs, as shown in Figure 3a [39]. They used mixtures of achiral cyclohexane derivative bearing three mesogens based on biphenylbenzoate and cyclohexane derivative **2** bearin a chiral mesogen. The mixture formed N* glassy state and stable LC thin films wer produced. An oligofluorene dye **3** was dissolved in the N* matrices. Within the reflectio band, CP PL was obtained with the g_e value of 1.8 in a 35 μm-thick film at the maximur However, the width of the reflection band was only 60 nm and high g_e value was obtaine in 20 nm (Figure 3b). As indicated in Equation (2), the band-width is proportional to th birefringence. This narrow region for the CP light emission was attributed to the sma birefringence of the matrix LCs. The concentration of the π-conjugated chromophores wa less than 1 mol %, and the LC glassy films were electrically insulative.

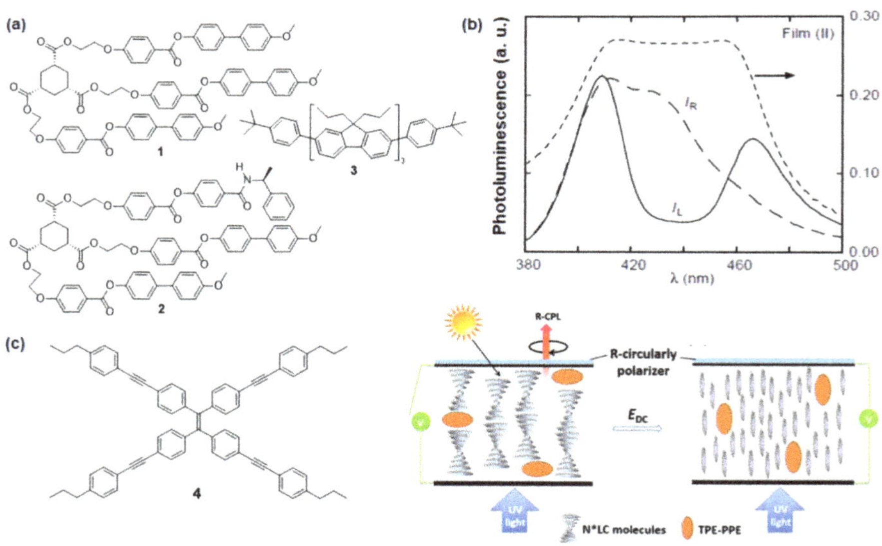

Figure 3. (**a**) Molecular structures of the components **1–3** of glassy N* films emitting CP fluorescence. (**b**) CP fluorescence and absorption spectra from the glassy film. I_R and I_L denote the spectrum of right-handed left-handed PL, respectively. The spectra were reproduced from Reference [39] with the permission from Springer and Nature. (**c**) CP light device using a N* liquid crystal doped with a dye **4** exhibiting AIE activity. The images were reproduced from Reference [40] with the permission from John Wiley and Sons.

Zhao et al. demonstrated an LC display using N* LC dissolving dye **4**, which indicate aggregation-induced emission (AIE) (Figure 3b) [40]. The photoluminescence spectrum c the dye covered visible light (400~600 nm) and CP light emission was obtained in the regio of the photonic band with the g_e value of 0.4 at best. The switching between non-polarize and CP light emission by the application of a DC voltage. They proposed a display that used reflection of sunlight in the daytime and emission from the dye excited by UV light at night.

The introduction of an extended π-conjugated unit is effective to increase the birefringence of the N* liquid crystals as well as π-orbital overlap to promote electronic charge carrier transport. O'Neill and Kelly synthesized fluorene-based LC semiconductors **5** and **6** bearing chiral side chains, as shown in Figure 4a [41]. These materials formed glassy N* film at room temperature and π-conjugated units aggregated in the glassy N* phase with high intensity. These molecules had an extended π-conjugated unit and high hole mobility exceeding 10^{-4} $cm^2V^{-1}s^{-1}$ at room temperature. Moreover, the long π-conjugated system resulted in high birefringence and the width of the reflection band exceeded 300 nm covering a wide region of visible light. The g_e value of CP PL exceeded 1.5 from 457 to 561 nm in a sample with a thickness of 3 μm. The reflection band can be tuned by mixing compound **5** with compound **6**.

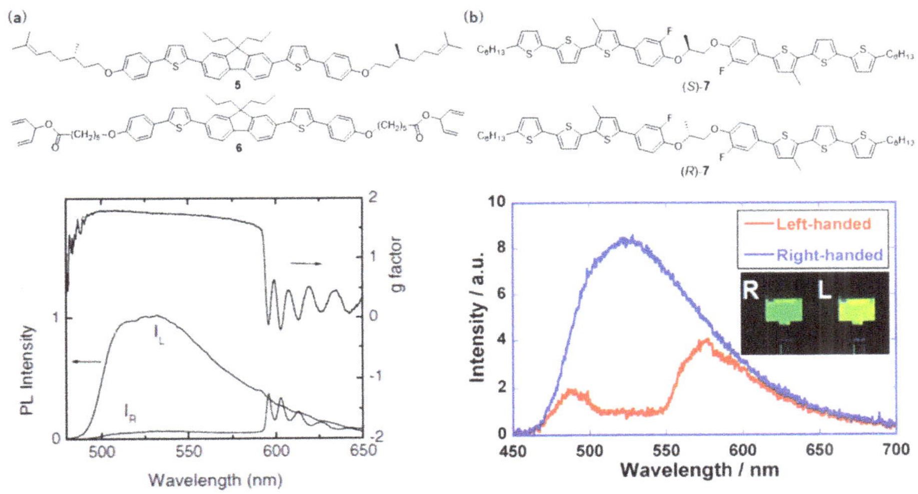

Figure 4. CP fluorescence from non-diluted π-conjugated N* liquid crystals. (**a**) Fluorene derivatives, CP PL spectra, and g factor as a function of wavelength. The spectra were reproduced from [41] with the permission from John Wiley and Sons. (**b**) Phenylterthiophene derivatives and CP PL spectra. The inset is a photograph of the LC cell emitting CP light through a CP filter. The spectra were reproduced from Reference [42] with the permission from the Royal Society of Chemistry.

Hamamoto and Funahashi synthesized dimeric N* LC based on phenylterthiophene skeleton (Figure 4b) [42]. Due to the small volume of the side chain and chiral moieties, the density of the π-conjugated chromophores was increased further. The LC **7** exhibited an N* phase at room temperature and ambipolar electronic charge carrier transport. Due to the extended π-conjugated system, the width of the reflection band was 100 nm. Within the reflection band, the g_e value of the CP PL reached 1.5 in a cell with a thickness of 9 μm. Reflection band could be red-shifted by mixing (*R*)- and (*S*)-**7**. The N* phase of LC **7** was fluidic at an elevated temperature and the switching between the CP and non-polarized states by the application of an electric field.

2.2. Circularly Polarized Electroluminescence from Twisted Electronic Systems

CP EL has been confirmed in the devices with light-emitting layers, including chiral rare-earth metal complexes, chiral conjugated polymers, and chiral low-molecular-weight dyes. In the cases of chiral metal complexes and low-molecular weight-dyes, the g_e values of electroluminescence were in the order of 10^{-3} to 10^{-2}. Exceptionally, chiral Pt complex works as effective CP light emitters in a few EL devices, indicating the g_e value of 0.3 most. Chiral low molecular weight dye exhibiting thermally activated delayed fluoresce has also been designed, but the g_e values are less than 0.1 at the present stage [43–46].

Conjugated polymers bearing chiral alkyl chains form twisted aggregates by soluti processes. A few polymers exhibit the N* order in the thin film states. This helical struct of π-conjugated aggregates enhances circularly polarization of the PL and EL.

The first CP EL device was reported by Meijer and coworkers in 1997 [47], as sho in Figure 5a. They synthesized poly(*p*-phenylenevinylene) derivative **8** bearing chiral al chains and fabricated EL devices, including light-emitting layer of the conjugated polym with a thickness of 300 nm by a spin-coating method. The g_e value was 1.7×10^{-3} 600 nm. They improved the g_e value to 0.35 in an EL device with a thickness of 70 r using a nonafluorene derivative bearing chiral alkyl chains [48].

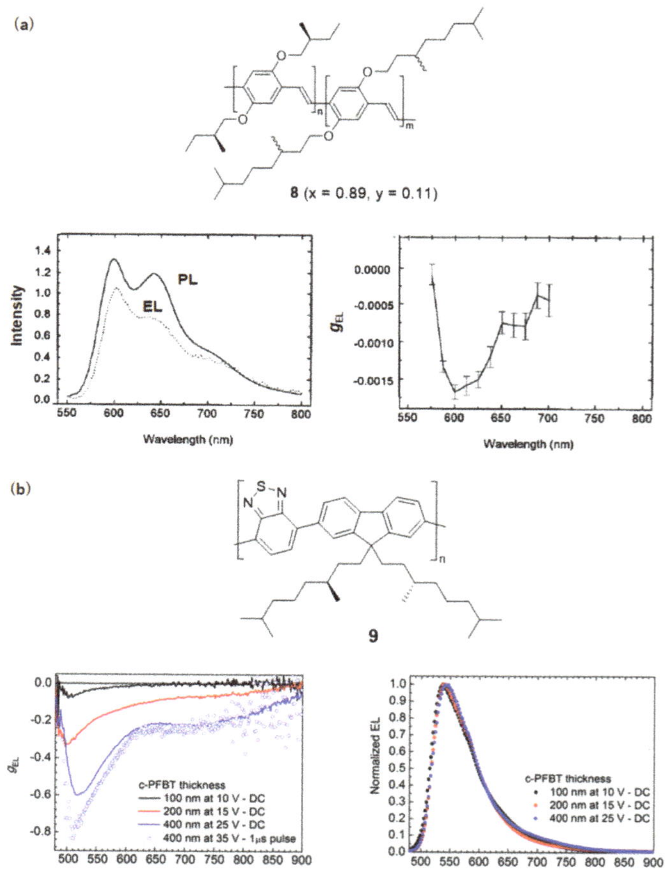

Figure 5. (**a**) Chiral poly(*p*-phenylenevinylene) derivative as well as EL and PL spectra. The g_e val for EL as a function of the wavelength is also indicated. The spectra were reproduced from [47] w the permission from the American Chemical Society. (**b**) Chiral polyfluorene and the g_e value of t EL from samples with various thicknesses as a function of the wavelength. EL spectra from samp with various thicknesses are also indicated. The graphs were reproduced from Reference [49] w the permission from the American Chemical Society (https://pubs.acs.org/doi/10.1021/acsna 7b07390 accessed on 1 January 2021).

In 2017, Nuzzo and coworkers reported CP EL using a poly(fluorene-co-benzothiadiazole) derivative **9** bearing chiral alkyl side chains [49], as shown in Figure 5b. The g_e value increased with the increase in the thickness of the light emitting layer, to be 0.8 at 500 nm with the thickness of 400 nm under pulse DC bias application. This result indicated that the generation of CP light was related to the helical structure of the conjugated polymer layer and relatively thick light emitting layer should be required for a high g_e value.

CP EL was observed from the devices with light emitting layers using achiral conjugated polymer **10** doped with chiral inactive compound **11**, as shown in Figure 6a. The CP EL devices with the g_e value approaching 1 and the luminance in the order of 10^3 cdm^{-2} were fabricated [50]. The g_e value of EL is higher than that of PL and depended on the film thickness and twist angle of the conjugated polymer orientation. As shown in Figure 6b, light was emitted from the whole area of the sample in PL, while light was emitted from a specific area in EL. Additionally, the g_e value of EL changed depending on the distance between the emission zone and the cathode, indicating the influence of the reflection of the CPL at the cathode on the g_e value [51,52]. In order to obtain high quality CP EL, reflection from the cathode, as well as twisted angle of the conjugated polymer aggregation, should be considered in the CP EL device design.

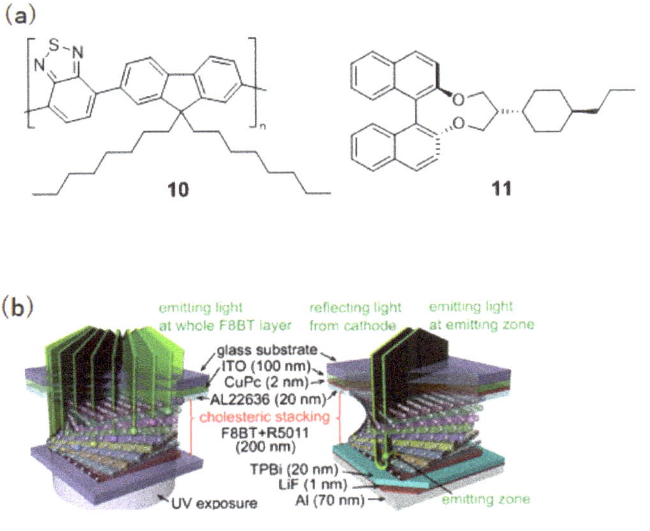

Figure 6. (**a**) Molecular structures of conjugated polymer **10** and chiral dopant **11** for CP EL devices. (**b**) Schematic illustration for CP PL (left) and CP EL (right) affected by the twist angle of the conjugated polymers, the position of light emission zone and reflection from the cathode. The image was reproduced from Reference [51] with the permission from John Wiley and Sons.

3. Liquid Crystalline Electronic Systems with Ferroelectricity

Conventional ferroelectrics have large bandgaps, resulting in their low charge carrier densities. Therefore, they are electrically insulative. However, electrical conductivity can be induced if electronic charge carriers are injected into the ferroelectrics. In such a situation, the electronic charge carrier transport can be coupled with the electrical polarization originating from the ferroelectricity. The bulk photovoltaic effect observed in lithium niobate and bismuth ferrite is attributed to the coupling between the transport of photogenerated charge carriers and the internal electric field produced by the spontaneous polarization [53].

3.1. Bulk Photovoltaic Effect in Ferroelectric Solids

Figure 7a shows a schematic illustration of a photovoltaic effect based on a p-n junction. The conventional photovoltaic effect is driven by local electric fields formed at junctions between p- and n-type semiconductors in a case of a p-n junction, or a metal and a semiconductor in a case of a Schottky junction. The built-in potential is determined by the difference between chemical potentials of semiconductors and metals [54]. Therefore, a built-in potential does not exceed a bandgap of a semiconductor. Open-circuit voltages of solar cells based on inorganic and organic solar cells are typically around 1 V [55]. For solar cells based on the p-n junction, series connection of solar cells is required for generation of a high voltage larger than the bandgap. Tandem structures of photovoltaic components using different active materials are also studied for the increase in the open circuit voltages and conversion efficiencies of inorganic and organic solar cells [56,57].

In contrast to the conventional photovoltaic effect, a larger open-circuit voltage can be generated in an anomalous or bulk photovoltaic effect because the driving force of the photovoltaic effect is a strong electric field formed in the bulk of the materials because the open circuit voltage is proportionate to the electric polarization of the material, as shown in Figure 7b [53]. Crystalline films of lithium niobate and barium titanate exhibit the bulk photovoltaic effect in which the open circuit voltages exceed their bandgaps [58,59]. However, the photocurrents were very low, resulting in low power conversion efficiency because of their large bandgaps and low conductivities. Recently, the bulk photovoltaic effect in bismuth ferrite has been studied. Bismuth ferrite exhibits a ferroelectric phase at room temperature and has absorption edge around 560 nm. Choi et al. observed the inversion of the polarity of rectification and photocurrent by the change of poling bias polarity [60]. The significance of polarization boundaries is recognized for the generation of high open-circuit voltages in the bulk photovoltaic effect, as shown in Figure 7c [61].

For organic electronic materials, the examples of the bulk photovoltaic effect have been quite limited. First of all, the variety of organic ferroelectrics is limited compared to inorganic materials [62]. A classical example is single crystals of tartaric acid salts. Crystals in which hydrogen-bonding networks are formed, such as croconic acid, are extensively studied [63]. Poly(vinylidene fluoride) films are typical ferroelectric polymers and applications to sensors and memories are investigated [64]. Ferroelectric liquid crystals are studied for display applications [65]. Above-mentioned organic ferroelectric materials bear no or small π-conjugated systems and fundamentally electrical insulators. In these materials, photovoltaic effect could not be expected because of low electrical conductivity and no absorption in UV-visible light region. For this purpose, extended π-conjugated systems, which absorb UV-visible light and contribute to electronic charge carrier transport, should be built in the polar molecules.

Among organic systems, the first case of bulk photovoltaic effect was observed in films of dye-doped poly(vinylidene fluoride) **12** (Figure 8a) by Sasabe and coworkers [66]. Poly(vinylidene fluoride) is a typical ferroelectric polymer exhibiting high spontaneous polarization around 5 μCcm^{-2} in the β-form. However, it is an electrical insulator and has no absorption bands in the visible and near UV region. Therefore, acridine orange absorbing visible light was doped in the polymer films. In the poled films, photocurrent response and pyroelectric effect were observed under open-circuit conditions, as shown in Figure 8a. A high open-circuit voltage in the order of 10^4 was obtained, but the photocurrent was too small, resulting in a low conversion efficiency of 0.25%. Moreover, the dye molecules worked as carrier traps to decrease the photocurrent.

Tasaka and coworkers reported a bulk photovoltaic effect of which polarity could be inverted by poling DC bias applications in a polycrystalline thin film of triphenylene hexacarboxylic ester [67]. Figure 8b shows a molecular structure of the ferroelectric triphenylene derivative and the photocurrent response in the thin film state. In crystalline thin films of 2,3,6,7,10,11-hexakis(4-octyloxybenzoyloxy)triphenylene **13**, a macroscopic electric polarization is induced by the orientation of dipole moments of the six carbonyl groups connecting to a triphenylene core, as shown in Figure 8b. A hole transport proceeded through

π-stacks consisting of triphenylene units. In the crystal phase of this compound, the spontaneous polarization was 200 $nCcm^{-2}$. When a UV light of the intensity of 10 $mWcm^{-2}$ was illuminated, the photocurrent density reached 5 $nAcm^{-2}$, as shown in Figure 8b. The power conversion efficiency was 0.5% and the internal electric field should be deduced to be 1.4 $MVcm^{-1}$. This material has π-stacking structures, which is favorable for an efficient electronic charge carrier transport. However, the carrier generation efficiency should not be sufficiently high, resulting in low conversion efficiency.

Nakamura and coworkers confirmed a shift current in ferroelectric charge transfer complex of tetrathiafulvalene **14** and chloranil **15**, which formed a non-centrosymmetric crystalline structure (Figure 9) [68]. In the single crystals, ferroelectricity was derived from charge transfer between the electron-acceptor and donor. The spontaneous polarization exceeded 5 $mCcm^{-2}$ below 70 K. By visible light illumination, directional photocurrent was observed and the polarity could be inverted by the change in the polarity of the poling DC bias. However, the crystals exhibited a ferroelectric phase only below 80 K and consequently this bulk photovoltaic effect disappeared at room temperature.

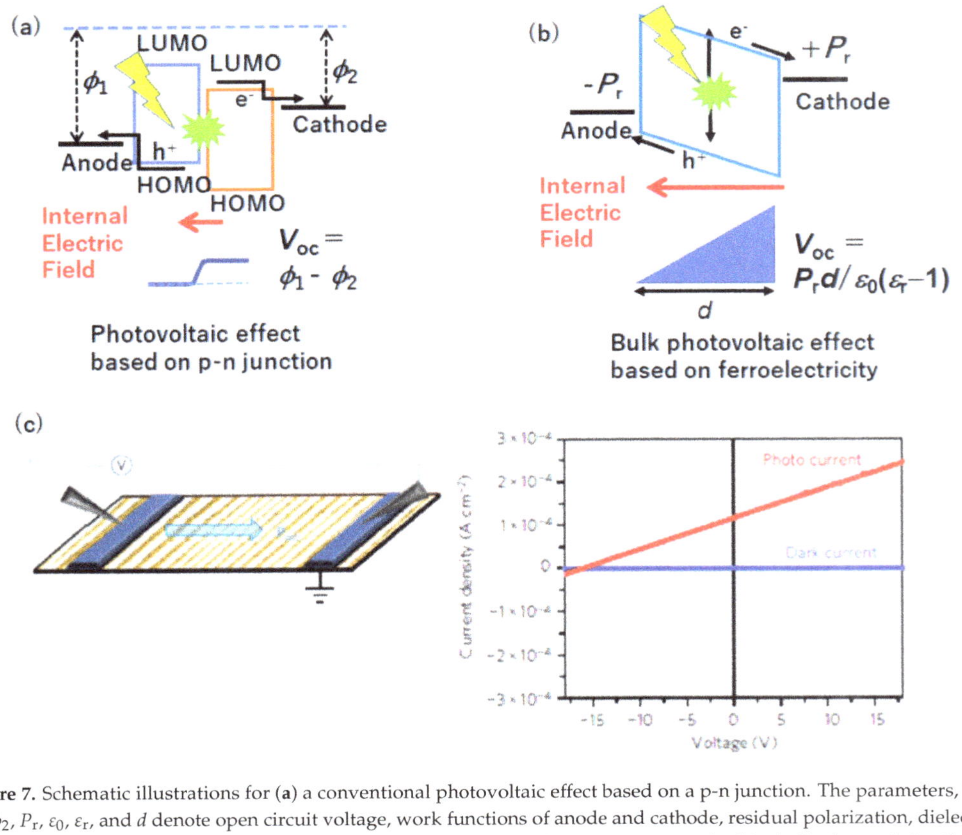

Figure 7. Schematic illustrations for (**a**) a conventional photovoltaic effect based on a p-n junction. The parameters, V_{oc}, ϕ_1, ϕ_2, P_r, ε_0, ε_r, and d denote open circuit voltage, work functions of anode and cathode, residual polarization, dielectric constants of vacuum, relative permittivity of the sample, and sample thickness, respectively. (**b**) a bulk photovoltaic effect in a ferroelectric. (**c**) An example of the bulk photovoltaic effect in bismuth ferrite. A schematic illustration for a measurement of the planar sample containing polarization boundaries and the current-voltage characteristics. The image and graph were reproduced from reference [61] with the permission from Springer Nature.

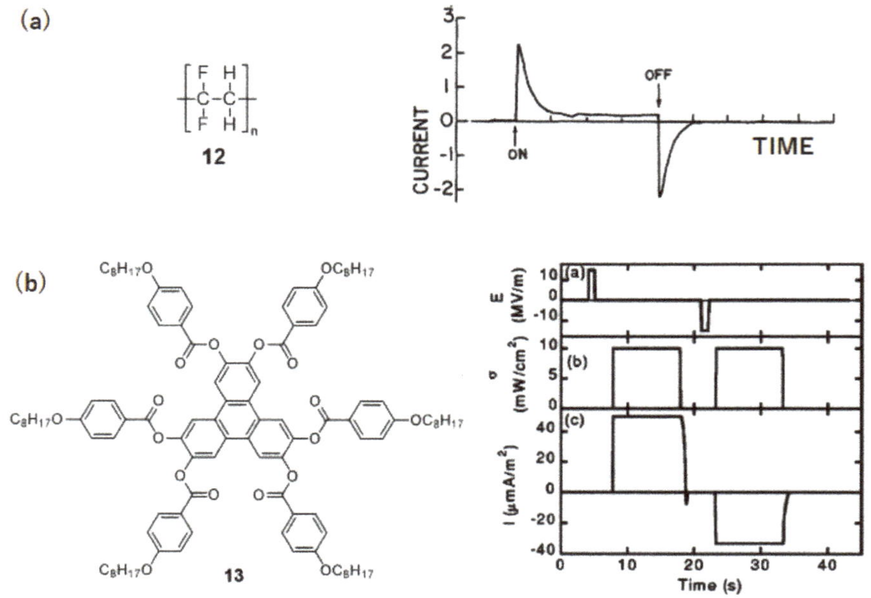

Figure 8. (**a**) Molecular structure of poly(vinylidene fluoride) **12** and the photocurrent response of the poled thin film under a short-circuit condition. The graph was reproduced from Reference [66] with the permission from Springer Nature. (**b**) Molecular structure of 2,3,6,7,10,11-hexakis(4-octyloxybenzoyloxy)triphenylene **13** and the photocurrent response under a short-circuit condition. The graph was reproduced from Reference [67] with the permission from the American Physical Society.

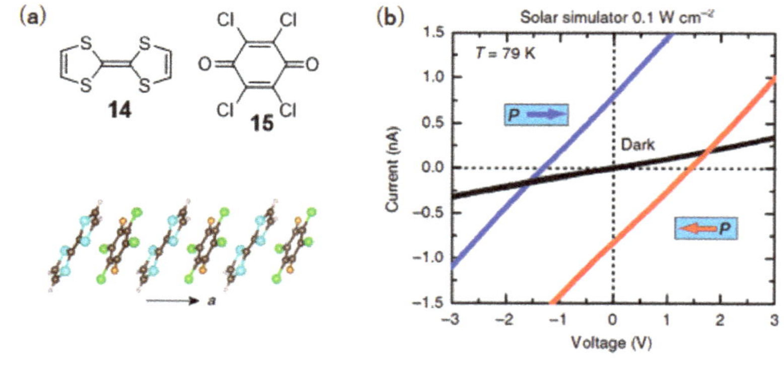

Figure 9. (**a**) Molecular structures of a charge transfer complex between tetrathiafulvalene **14** and chloranil **15** and the crystal structure of the charge transfer complex. (**b**) Current-voltage characteristics of the charge transfer complex at 79 K. The image and graph were reproduced from Reference [68] with the permission from Springer Nature.

3.2. Bulk Photovoltaic Effect in Liquid Crystalline Systems

3.2.1. Ferroelectric Liquid Crystals

Figure 10 shows molecular structures of typical Ferroelectric liquid crystals (FLCs). FLCs were discovered by Meier and coworkers in 1975 [69] and applied to display devices by Clark and Lagerwall in 1980 [70]. The first type of FLC molecules consisted of a rod-like rigid core and a chiral alkyl chain, which broke a centrosymmetry of the mesomorphic system. The first example of FLC is compound **16**. The FLC phases have a layer structure in which the molecules are tilted to the layer normal. The typical FLC phase is a chiral smectic C (SmC*) phase, as shown in Figure 1b.

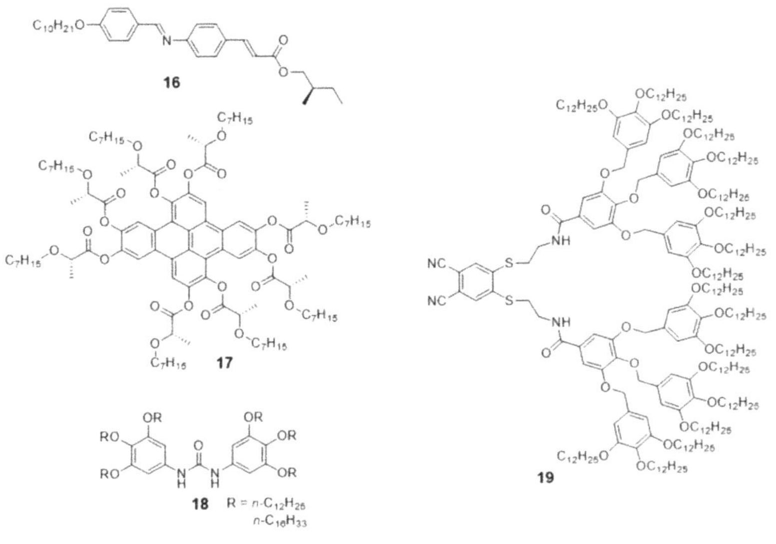

Figure 10. Molecular structures of ferroelectric liquid crystals.

The second type of the FLC phase is columnar phases. Bock and Helfrich synthesized and dibenzopyrene derivative **17** bearing chiral alkyl chains to exhibit ferroelectricity [71]. As observed in the SmC* phase, non-centrosymmetry induced by tilted normals of the π-conjugated disks to the columnar axes as well as molecular chirality was the origin of the ferroelectricity. The spontaneous polarization was induced perpendicular to the columnar axes. In contrast, Kishikawa and coworkers synthesized achiral diphenyl urea derivatives **18** bearing long alkyl chains, which formed a polar columnar phase in 2005 [72]. In the columnar phase, the origin of the electrical polarization was one-dimensional hydrogen-bonding series to generate the polarization parallel to the columnar axes. Polarization inversion was caused by the exchange of the hydrogen bonding between urea molecules.

In 2012, Miyajima et al. achieved the construction of ferroelectric columnar phase by an approach from supramolecular chemistry [73]. They synthesized achiral phthalonitrile derivative **19** bearing amide moieties and alkyl side chains. Four molecules of compound **19** form an umbrella-like aggregate via hydrogen bonding between amide moieties and they stack in a one-dimensional manner to form columnar aggregates. The cyano groups of phthalonitrile units generated a macroscopic electrical polarization parallel to the columnar axes. They confirmed the ferroelectricity by a polarization inversion current technique and second harmonic generation (SHG) measurement.

In the abovementioned studies, the LC molecules did not comprise extended π-conjugated systems and the FLC phases were electrically insulative. For the coupling of the ferroelectricity with electronic functions, extended π-conjugated units and polar moieties should be integrated in the molecules and they should be organized properly in the supramolecular structures.

3.2.2. Chiral LC Systems

Funahashi and coworkers have paid attention to chiral supramolecular systems. They synthesized FLCs based on a phenylterthiophene skeleton [74]. A chiral alkyl chain and a fluorophenyl unit were connected to a terthiophene core, which exhibited good carrier transport characteristics in smectic phases. Phenylterthiophene derivatives **20–22** (Figure 11a) exhibited a SmC* and SmG* phases and the hole mobilities were in the order of 10^{-4} $cm^2V^{-1}s^{-1}$ in the SmC* phase and 10^{-2} $cm^2V^{-1}s^{-1}$ in the SmG* phase. The spontaneous polarization reached 100 $nCcm^{-2}$ in the SmC* phase.

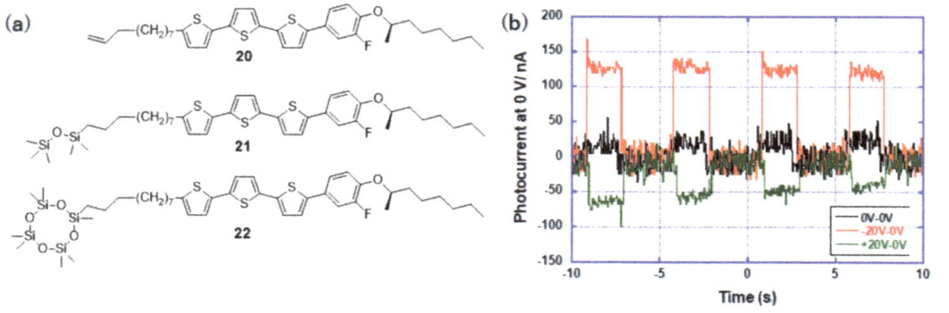

Figure 11. (**a**) Molecular structure of ferroelectric LC semiconductors **20–22**. (**b**) Photocurrent response in the SmC* phase of compound **20** for UV light illumination under a short circuit condition. Black, red, and green lines denote a non-polarized state, polarized state with the illuminated electrode charged positively, and polarized state with that charged negatively, respectively. The graph was reproduced from Reference [74] with the permission from the Royal Society of Chemistry.

Figure 11b shows photocurrent response for UV light illumination in the SmC* phase of compound **20** under a short circuit condition. For UV light illumination, a photocurrent was induced under the zero bias after the application of a poling DC bias. The polarity of the photocurrent was contrary to that of the poling bias and it was inversed by the change in the polarity of the poling bias. The same photovoltaic behaviors were observed in the SmC* phase of the other phenylterthiophene derivatives and the photocurrent response was proportional to the spontaneous polarization in the SmC* phase [75].

The photovoltaic response of which polarization is contrary to the poling bias prior to light illumination can be induced by an electrical polarization of ionic impurities. The origin of the photovoltaic effect was confirmed by the comparison of photovoltaic behaviors between the FLC samples with various enantiomer purities [76]. As shown in Figure 12a, the photocurrent in the SmC* phase of compound **20** increased with an increase in the enantiomer purity. The photovoltaic effect disappeared in the racemic mixture. This result indicates that the photovoltaic effect should be attributed to the symmetry breaking by molecular chirality in the SmC* phase, but not to contamination of ions.

These phenylterthiophene-based FLCs exhibit chiral ordered smectic phases below the SmC* phase. If a sandwich type sample of the FLCs is cooled from the SmC* phase to the ordered smectic phase under the DC bias application, the macroscopic electric polarization in the SmC* phase is immobilized in the ordered smectic phase. The increase of the carrier mobility in the ordered smectic phase compared to that in the SmC* phase results in the enhancement of the photovoltaic effect in the ordered smectic phase, as shown in Figure 12b [77].

Funahashi and coworkers synthesized a double chiral FLC **23** based on bisfluorophenylterthiophene bearing lactic ester moieties, as shown in Figure 13a [78]. This compound exhibited a chiral smectic A (SmA*) phase below which an ordered smectic phase appeared. When a thin film sample was cooled from the SmA* phase to the ordered smectic phase under the application of a DC bias, the macroscopic electrical polarization was induced in the ordered smectic phase. Figure 13b shows a current-voltage characteristic for UV light illumination in the polarized state of the chiral ordered smectic phase of compound **23**. Due to the internal electric field produced by the macroscopic polarization, the bulk photovoltaic effect was confirmed in the ordered smectic phase of this compound with an open-circuit voltage of 0.8 V. Moreover, the electrical polarization in the ordered smectic phase lowered the injection barrier from the cathode and anode to the FLC layer, resulting in electroluminescence. From uniaxially aligned samples, linearly polarized electroluminescence was obtained and the polarized plane of the electroluminescence was rotated by 90 degrees. by the inversion of the poling bias polarity in the cooling process from the SmA* phase to the ordered smectic phase, as shown in Figure 13c.

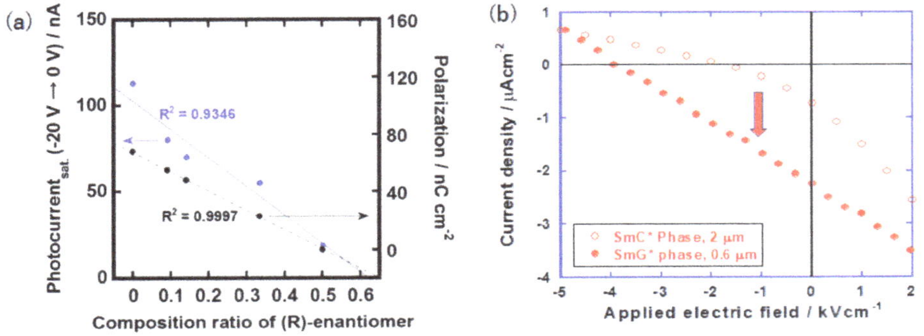

Figure 12. (**a**) Photocurrent under a short circuit condition in the SmC* phase of compound **20** as a function of the enantiomer purity. The graph was reproduced from Reference [76] with permission from the PCCP Owner Societies. (**b**) Current-voltage characteristics in the SmC* and SmG* phases of compound **20**. The graph was reproduced from Reference [77] with the permission from Elsevier.

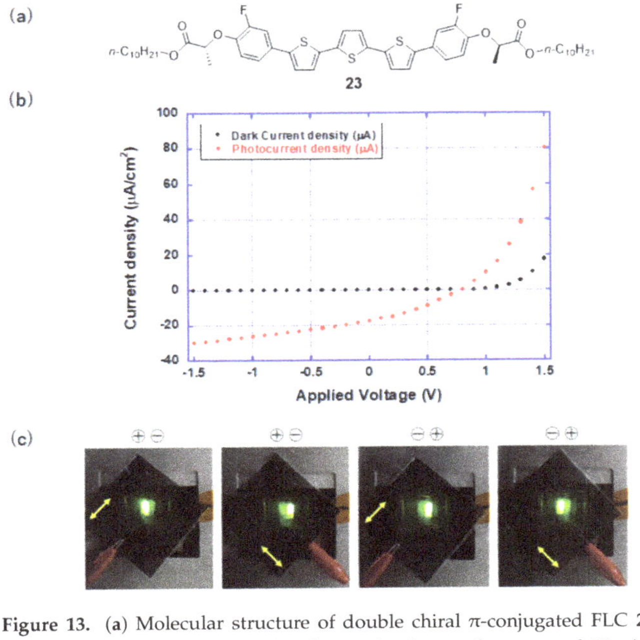

Figure 13. (**a**) Molecular structure of double chiral π-conjugated FLC **23**. (**b**) Current-voltage characteristics in the chiral ordered smectic phases of compound **23**. (**c**) Electric-field-induced rotation of the polarized plane of the electroluminescence from the polarized ordered smectic phase of compound **23**. The photographs and graph were reproduced from Reference [78] with the permission from the Royal Society of Chemistry.

The bulk photovoltaic effect in the chiral ordered smectic phase of compound **20** was enhanced by doping of fullerene derivative **24**. With the concentration of 13 mol % of compound **24** in compound **20**, fullerene-rich domains were formed in the chiral ordered phase. The photovoltaic effect was not so remarkably enhanced because the domain size was much larger than the exciton diffusion length. However, the spectral sensitivity was extended to the visible light area. The polarity of the effect could be inverted by the change of the poling bias during the cooling process from the SmC* phase to the ordered smectic phase, as shown in Figure 14a.

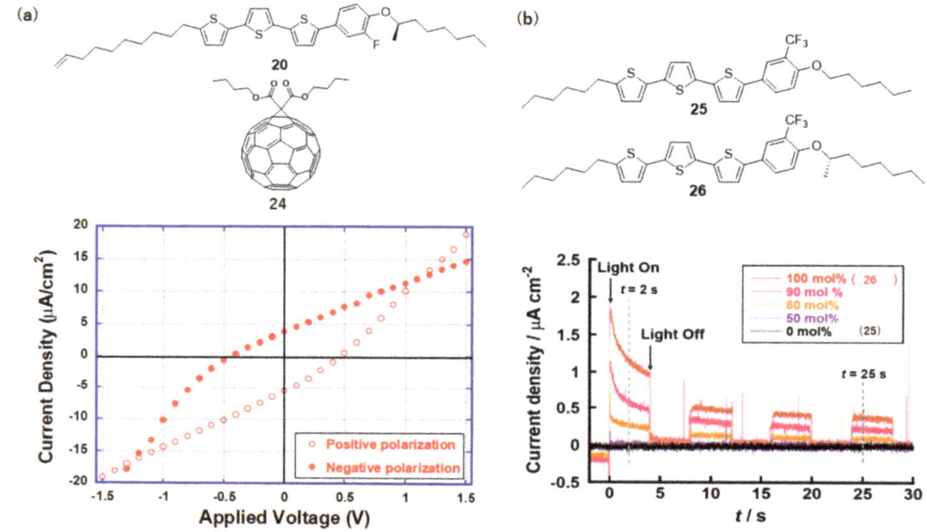

Figure 14. (**a**) Current-voltage characteristic in the polarized chiral ordered smectic phase of compound **20** doped with fullerene derivative **24** (13 mol %) for white light illumination. The graph was reproduced from Reference [79] from the permission of Elsevier. (**b**) Photocurrent response in the SmC* phase of the mixtures of compounds **25** and **26** under a short circuit condition. The graph was reproduced from Reference [80] from the permission from the American Chemical Society.

This bulk photovoltaic effect was also observed in binary systems consisting of a chiral and achiral FLCs [79]. The chirality induced by a small amount of chiral dopant can break the centrosymmetry of the LC electronic system, resulting in the photovoltaic effect in the ferroelectric LC phase [80]. As shown in Figure 14b, the photovoltaic effect was induced by doping of chiral compound **26** to achiral LC **25**.

3.2.3. Achiral LC Systems

A few ferroelectric columnar LCs comprising extended π-conjugated units have been synthesized. In the studies, polar moieties and extended π-conjugated cores are integrated artfully in one-dimensional columnar structures.

Akutagawa and coworkers reported ferroelectric columnar LC **27** consisting of pyrene units and amide moieties to form one-dimensional hydrogen bonding network (Figure 15a) [81]. The electronic charge carrier transport proceeded through π-stacks consisting of pyrene cores while the hydrogen bonding networks between amide moieties induced ferroelectricity. They confirmed a shift of the current-voltage characteristics by the polarity inversion of a DC poling bias, as shown in Figure 15a, although no photovoltaic effect was observed in the ferroelectric phase.

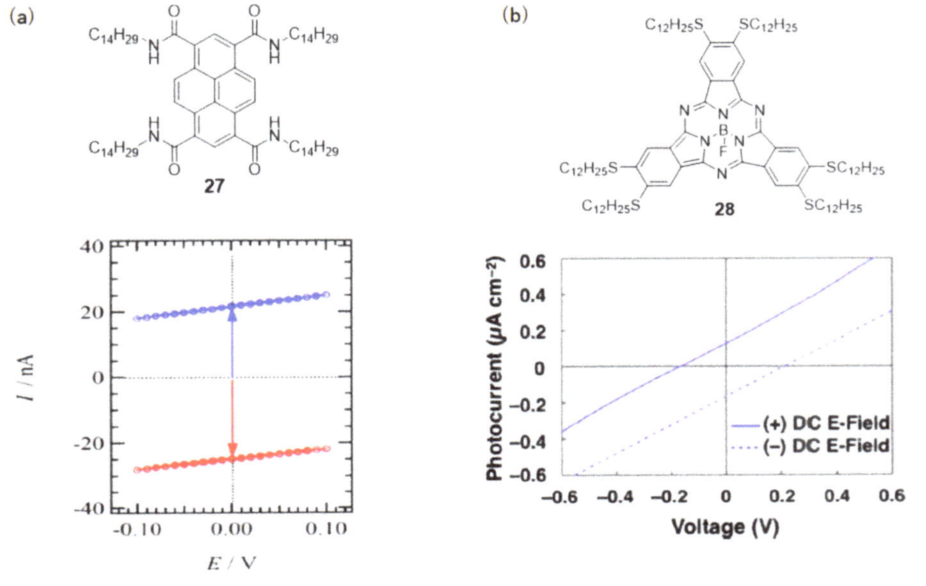

Figure 15. (**a**) Columnar FLC **27** based on a pyrene moiety and the current-voltage characteristic in the ferroelectric columnar phase. The graph was reproduced from Reference [81] with the permission of the American Chemical Society. (**b**) Ferroelectric subphthalocyanine derivative **28** and the current voltage characteristic in the polarized columnar phase of compound **28** for visible light illumination. The graph was reproduced from Reference [82] with the permission of the American Chemical Society.

Miyajima and coworkers synthesized subphthalocyanine derivative **28** exhibiting a ferroelectric columnar phase (Figure 15b) [82]. This subphthalocyanine core including boron atom was non-planar and stacked to form polarized columnar aggregates. For visible light illumination, photoconductive behavior was observed in the polarized columnar phase. The polarity of the photocurrent could be inverted by the change of the polling bias polarity and the open-circuit voltage depended on the sample thickness, as shown in Figure 15b, indicating that this photoconductivity should be attributed to a bulk photovoltaic effect.

4. Conclusions

Chiral liquid crystals based on extended π-conjugated units form the N* and SmC* phases, in which electronic function originated from the π-conjugated moieties couples with the molecular chirality to produce new optical and electronic functions. In the N* phase, which has twisted structures, CP PL has been observed within the wavelength range of the reflection band. Recently, CP EL has been confirmed from devices based on active layers consisting of chiral conjugated polymers with twisted structures induced by the molecular chirality. FLCs, consisting of phenylterthiophene skeleton and chiral alkyl side chains, indicate a bulk photovoltaic effect, which is driven by spontaneous polarization. The bulk photovoltaic effect is enhanced in the chiral ordered smectic phases and by the formation of bulk heterojunctions with fullerene derivatives. This bulk photovoltaic effect has also been observed in achiral ferroelectric columnar LCs, which has extended π-conjugated units.

Funding: This study was financially supported by a Grant-in-Aid for Scientific Research on Innovative Areas (Element-Block Polymers, no. 15H00753) from the Ministry of Education, Culture, Sports, Science and Technology (MEXT), a Grant-in-Aid for Scientific Research (B) (no.15H03797, 21H01904) from the Japan Society for the Promotion of Science (JSPS), the Ogasawara Foundation for the Promotion of Science & Engineering, the SEI Group CSR Foundation, the Asahi Glass Foundation, the Salt Science Research Foundation (no. 1715), Iketani Science and Technology Foundation, and the TEPCO Memorial Foundation. This research was also supported by 'Nanotechnology Platform Program' of the Ministry of Education, Culture, Sports, Science and Technology (MEXT), Japan (Grant No. JPMXP09F19GA0004).

Institutional Review Board Statement: Not applicable.

Informed Consent Statement: Not applicable.

Data Availability Statement: Not applicable.

Acknowledgments: The author thanks Akinari Sonoda (AIST, Shikoku) for NMR measurements, Kusunose (Kagawa Univ.) for DSC measurements, and Tomohiko Ishii (Kagawa Univ.) for X-ray diffraction analysis.

Conflicts of Interest: The author declares no conflict of interest.

References

1. De Gennes, G.P.; Prost, J. *The Physics of Liquid Crystals*, 2nd ed.; Clarendon Press: Oxford, UK, 1997.
2. Lagerwall, S.T. *Ferroelectric and Antiferroelectric Liquid Crystals*; Wiley-VCH: Weinheim, Germany, 1999.
3. Kikuchi, H. Liquid Crystalline Blue Phases. *Struct. Bond.* **2007**, *128*, 99–117.
4. Renn, S.R.; Lubensky, T.C. Abrikosov dislocation lattice in a model of the cholesteric–to–smectic-A transition. *Phys. Rev. A* **1988**, *38*, 2132–2147. [CrossRef]
5. Funahashi, M. Development of Liquid-Crystalline Semiconductors with High Carrier Mobilities and Their Application to Thin-film Transistors. *Polym. J.* **2009**, *41*, 459–469. [CrossRef]
6. Funahashi, M. Nanostructured liquid-crystalline semiconductors—A new approach to soft matter electronics. *J. Mater. Chem. C* **2014**, *2*, 7451–7459. [CrossRef]
7. Funahashi, M. Solution-processable electronic and redox-active liquid crystals based on the design of side chains. *Flex. Print. Electron.* **2020**, *5*, 043001. [CrossRef]
8. O'Neill, M.; Kelly, S.M. Ordered Materials for Organic Electronics and Photonics. *Adv. Mater.* **2010**, *23*, 566–584. [CrossRef]
9. Funahashi, M.; Sonoda, A. Electron transport characteristics in nano-segregated columnar phases of perylene tetracarboxylic bisimide derivatives bearing oligosiloxane chains. *Phys. Chem. Chem. Phys.* **2014**, *16*, 7754–7763. [CrossRef] [PubMed]
10. Funahashi, M.; Ishii, T.; Sonoda, A. Temperature-Independent Hole Mobility of a Smectic Liquid-Crystalline Semiconductor based on Band-Like Conduction. *ChemPhysChem* **2013**, *14*, 2750–2758. [CrossRef]
11. Kato, T.; Yoshio, M.; Ichikawa, T.; Soberats, B.; Ohno, H.; Funahashi, M. Transport of ions and electrons in nanostructured liquid crystals. *Nat. Rev. Mater.* **2017**, *2*, 17001. [CrossRef]
12. Adam, D.; Closs, F.; Frey, T.; Funhoff, D.; Haarer, D.; Ringsdorf, H.; Schuhmacher, P.; Siemensmeyer, K. Transient photoconductivity in a discotic liquid crystal. *Phys. Rev. Lett.* **1993**, *70*, 457–460. [CrossRef]
13. van de Craats, A.M.; Warman, J.M.; Fechtenkotter, A.; Brand, J.B.; Harbison, M.A.; Mullen, K. Record Charge Carrier Mobility in a Room-Temperature Discotic Liquid-Crystalline Derivative of Hexabenzocoronene. *Adv. Mater.* **1999**, *11*, 1469–1472. [CrossRef]
14. Ban, K.; Nishikawa, K.; Ohta, K.; van de Craats, A.M.; Warman, J.M.; Yamamoto, I.; Shirai, H. Discotic liquid crystals of transition metal complexes 29: Mesomorphism and charge transport properties of alkylthiosubstituted phthalocyanine ra-re-earth metal sandwich complexes. *J. Mater. Chem.* **2001**, *11*, 321–331. [CrossRef]
15. Würthner, F.; Saha-Möller, C.R.; Fimmel, B.; Ogi, S.; Leowanawat, P.; Schmidt, D. Perylene Bisimide Dye Assemblies as Archetype Functional Supramolecular Materials. *Chem. Rev.* **2016**, *116*, 962–1052. [CrossRef] [PubMed]
16. Funahashi, M.; Sonoda, A. High electron mobility in a columnar phase of liquid-crystalline perylene tetracarboxylic bisimide bearing oligosiloxane chains. *J. Mater. Chem.* **2012**, *22*, 25190–25197. [CrossRef]
17. Funahashi, M.; Hanna, J.-I. Fast ambipolar carrier transport in smectic phases of phenylnaphthalene liquid crystal. *Appl. Phys. Lett.* **1997**, *71*, 602–604. [CrossRef]
18. Funahashi, M.; Hanna, J.-I. High Carrier Mobility up to 0.1 $cm^2\ V^{-1}\ s^{-1}$ at Ambient Temperatures in Thiophene-Based Smectic Liquid Crystals. *Adv. Mater.* **2005**, *17*, 594–598. [CrossRef]
19. Funahashi, M.; Kato, T. Design of Liquid Crystals: From George Gray's Nematic Molecules to Thiophene-based π-Conjugated Molecules. *Liq. Cryst.* **2015**, *42*, 909–917.
20. Zhang, H.; Shiino, S.; Shishido, A.; Kanazawa, A.; Tsutsumi, O.; Shiono, T.; Ikeda, T. A Thiophene Liquid Crystal as a Novel π-Conjugated Dye for Photo-Manipulation of Molecular Alignment. *Adv. Mater.* **2000**, *12*, 1336–1339. [CrossRef]

21. Matsui, A.; Funahashi, M.; Tsuji, T.; Kato, T. Hole Transport in Liquid-Crystalline Polymers with a Polysiloxane Backbone and a Phenylterthiophene Moiety in the Side Chain. *Chem. Eur. J.* **2010**, *16*, 13465–13472. [CrossRef]
22. Aldred, M.P.; Contoret, A.E.A.; Farrar, S.R.; Kelly, S.M.; Mathieson, D.; O'Neill, M.; Tsoi, W.C.; Vlachos, P. A Full-Color Electroluminescent Device and Patterned Photoalignment Using Light-Emitting Liquid Crystals. *Adv. Mater.* **2005**, *17*, 1368–1372. [CrossRef]
23. Funahashi, M.; Zhang, F.; Tamaoki, N. High ambipolar mobility in highly ordered smectic phase of dialkylphenylter-thiophene derivative that can be applied to solution-processed organic field effect transistors. *Adv. Mater.* **2007**, *19*, 353–358. [CrossRef]
24. Pisula, W.; Menon, A.; Stepputat, M.; Lieberwirth, I.; Kolb, U.; Tracz, A.; Sirringhaus, H.; Pakula, T.; Mullen, K. A Zone Cast-ing Technique for Device Fabrication of Field-Effect Transistors Based on Discotic Hexa-peri-Hexabenzocoronene. *Adv. Mater.* **2005**, *17*, 684–689. [CrossRef]
25. Van Breemen, A.J.J.M.; Herwig, P.T.; Chlon, C.H.T.; Sweelssen, J.; Schoo, H.F.M.; Setayesh, S.S.; Hardeman, W.M.; Martin, C.A.; De Leeuw, D.M.; Valeton, J.J.P.; et al. Large Area Liquid Crystal Monodomain Field-Effect Transistors. *J. Am. Chem. Soc.* **2006**, *128*, 2336–2345. [CrossRef]
26. Zhang, F.; Funahashi, M.; Tamaoki, N. High-performance thin film transistors from semiconducting liquid crystalline phases by solution processes. *Appl. Phys. Lett.* **2007**, *91*, 063515. [CrossRef]
27. Iino, H.; Usui, T.; Hanna, J.-I. Liquid crystals for organic thin-film transistors. *Nat. Commun.* **2015**, *6*, 6828. [CrossRef]
28. Hori, T.; Miyake, Y.; Yamasaki, N.; Yoshida, H.; Fujii, A.; Shimizu, Y.; Ozaki, M. Solution Processable Organic Solar Cell Based on Bulk Heterojunction Utilizing Phthalocyanine Derivative. *Appl. Phys. Express* **2010**, *3*, 101602. [CrossRef]
29. Shin, W.; Yasuda, T.; Watanabe, G.; Yang, Y.S.; Adachi, C. SelfOrganizing Mesomorphic Diketopyrrolo-pyrrole Deriva-tives for Efficient Solution-Processed Organic Solar Cells. *Chem. Mater.* **2013**, *25*, 2549. [CrossRef]
30. Schmidt-Mende, L.; Fechtenkötter, A.; Müllen, K.; Moons, E.; Friend, R.H.; MacKenzie, J.D. Self-Organized Discotic Liquid Crystals for High-Efficiency Organic Photovoltaics. *Science* **2001**, *293*, 1119–1122. [CrossRef]
31. Funahashi, M.; Tamaoki, N. Electronic Conduction in the Chiral Nematic Phase of an Oligothiophene Derivative. *ChemPhysChem* **2006**, *7*, 1193–1197. [CrossRef] [PubMed]
32. Funahashi, M.; Tamaoki, N. Effect of pretransitional organization in cholesteric phases of oligothiophene derivatives on their carrier transport characteristics. *Chem. Mater.* **2007**, *19*, 608–617. [CrossRef]
33. Woon, K.L.; Aldred, M.P.; Vlachos, P.; Mehl, G.H.; Stirner, T.; Kelly, S.M.; O'Neill, M. Electronic Charge Transport in Ex-tended Nematic Liquid Crystals. *Chem. Mater.* **2006**, *18*, 2311–2317. [CrossRef]
34. Tokunaga, K.; Takayashiki, Y.; Iino, H.; Hanna, J.-I. Electronic conduction in nematic phase of small molecules. *Phys. Rev. B* **2009**, *79*, 033201. [CrossRef]
35. Furumi, S.; Tamaoki, N. Glass-Forming Cholesteric Liquid Crystal Oligomers for New Tunable Solid-State Laser. *Adv. Mater.* **2010**, *22*, 886–891. [CrossRef]
36. Araoka, F.; Shin, K.-C.; Takanishi, Y.; Ishikawa, K.; Takezoe, H.; Zhu, Z.; Swager, T.M. How doping a cholesteric liquid crystal with polymeric dye improves an order parameter and makes possible low threshold lasing. *J. Appl. Phys.* **2003**, *94*, 279–283. [CrossRef]
37. Matsui, T.; Ozaki, R.; Funamoto, K.; Ozaki, M.; Yoshino, K. Flexible mirror-less laser based on a free-standing film of pho-topolymerized cholesteric liquid crystal. *Appl. Phys. Lett.* **2002**, *81*, 3741–3743. [CrossRef]
38. Finkelmann, H.; Kim, S.T.; Munoz, A.; Palffy-Muhoray, P.; Taheri, B. Tunable mirrorless lasing in cholesteric liquid crystal-line elastomers. *Adv. Mater.* **2001**, *13*, 1069–1072. [CrossRef]
39. Chen, S.H.; Katsis, D.; Schmid, A.W.; Mastrangelo, J.C.; Tsutsui, T.; Blanton, T.N. Circularly polarized light generated by photoexcitation of luminophores in glassy liquid-crystal films. *Nat. Cell Biol.* **1999**, *397*, 506–508. [CrossRef]
40. Zhao, D.; He, H.; Gu, X.; Guo, L.; Wong, K.S.; Lam, J.W.Y.; Tang, B.Z. Circularly Polarized Luminescence and a Reflec-tive-Photoluminescent Chiral Nematic Liquid Crystal Display Based on an Aggregation-Induced Emission Luminogen. *Adv. Opt. Mater.* **2016**, *4*, 534–539. [CrossRef]
41. Woon, K.L.; O'Neill, M.; Richards, G.J.; Aldred, M.P.; Kelly, S.M.; Fox, A.M. Highly circularly polarized photolumines-cence over a broad spectral range from a calamitic, hole-transporting, and chiral nematic glass and from an indirectly ex-cited dye. *Adv. Mater.* **2003**, *15*, 1555–1558. [CrossRef]
42. Hamamoto, T.; Funahashi, M. Circularly polarized light emission from a chiral nematic phenylterthiophene dimer exhib-iting ambipolar carrier transport. *J. Mater. Chem. C* **2015**, *3*, 6891–6900. [CrossRef]
43. Zhang, D.-W.; Li, M.; Chen, C.-F. Recent advances in circularly polarized electroluminescence based on organic light-emitting diodes. *Chem. Soc. Rev.* **2020**, *49*, 1331–1343. [CrossRef] [PubMed]
44. Zinna, F.; Giovanella, U.; Di Bari, L. Highly Circularly Polarized Electroluminescence from a Chiral Europium Complex. *Adv. Mater.* **2015**, *27*, 1791–1795. [CrossRef]
45. Brandt, J.R.; Wang, X.-H.; Yang, Y.; Campbell, A.J.; Fuchter, M.J. Circularly Polarized Phosphorescent Electroluminescence with a High Dissymmetry Factor from PHOLEDs Based on a Platinahelicene. *J. Am. Chem. Soc.* **2016**, *138*, 9743–9746. [CrossRef] [PubMed]
46. Feuillastre, S.; Pauton, M.; Gao, L.; Desmarchelier, A.; Riives, A.J.; Prim, D.; Tondelier, D.; Geffroy, B.; Muller, G.; Clavier, G.; et al. Design and Synthesis of New Circularly Polarized Thermally Activated Delayed Fluorescence Emitters. *J. Am. Chem. Soc.* **2016**, *138*, 3990–3993. [CrossRef] [PubMed]

47. Peeters, E.; Christiaans, M.P.T.; Janssen, R.A.J.; Schoo, H.F.M.; Dekkers, A.H.P.J.M.; Meijer, E.W. Circularly Polarized Electroluminescence from a Polymer Light-Emitting Diode. *J. Am. Chem. Soc.* **1997**, *119*, 9909–9910. [CrossRef]
48. Geng, Y.; Trajkovska, A.; Culligan, S.W.; Ou, J.J.; Chen, H.M.P.; Katsis, D.; Chen, S.H. Origin of Strong Chiroptical Activi-ties in Films of Nonafluorenes with a Varying Extent of Pendant Chirality. *J. Am. Chem. Soc.* **2003**, *125*, 14032–14038. [CrossRef]
49. Di Nuzzo, D.; Kulkarni, C.; Zhao, B.; Smolinsky, E.; Tassinari, F.; Meskers, S.C.J.; Naaman, R.; Meijer, E.W.; Friend, R.H. High Circular Polarization of Electroluminescence Achieved via Self-Assembly of a Light-Emitting Chiral Conjugated Polymer into Multidomain Cholesteric Films. *ACS Nano* **2017**, *11*, 12713–12722. [CrossRef]
50. Wan, L.; Wade, J.; Salerno, F.; Arteaga, O.; Laidlaw, B.; Wang, X.; Penfold, T.; Fuchter, M.J.; Campbell, A.J. Inverting the Handedness of Circularly Polarized Luminescence from Light-Emitting Polymers Using Film Thickness. *ACS Nano* **2019**, *13*, 8099–8105. [CrossRef]
51. Lee, D.-M.; Song, J.-W.; Lee, Y.-J.; Yu, C.-J.; Kim, J.-H. Control of Circularly Polarized Electroluminescence in Induced Twist Structure of Conjugate Polymer. *Adv. Mater.* **2017**, *29*, 1700907. [CrossRef]
52. Jung, J.-H.; Lee, D.-M.; Kim, J.-H.; Yu, C.-J. Circularly polarized electroluminescence by controlling the emission zone in a twisted mesogenic conjugate polymer. *J. Mater. Chem. C* **2017**, *6*, 726–730. [CrossRef]
53. Butler, K.T.; Frost, J.M.; Walsh, A. Ferroelectric materials for solar energy conversion: Photoferroics revisited. *Energy Environ. Sci.* **2015**, *8*, 838–848. [CrossRef]
54. Köhler, A.; Bässler, H. *Electronic Processes in Organic Semiconductors: An Introduction*; Wiley-VCH Verlag GmbH & Co. KGaA: Weinheim, Germany, 2015.
55. Elumalai, N.K.; Uddin, A. Open circuit voltage of organic solar cells: An in-depth review. *Energy Environ. Sci.* **2016**, *9*, 391–410. [CrossRef]
56. Werner, J.; Weng, C.-H.; Walter, A.; Fesquet, L.; Seif, J.P.; De Wolf, S.; Niesen, B.; Ballif, C. Efficient Monolithic Perov-skite/Silicon Tandem Solar Cell with Cell Area > 1 cm^2. *J. Phys. Chem. Lett.* **2016**, *7*, 161–166. [CrossRef] [PubMed]
57. Liu, W.; Li, S.; Huang, J.; Yang, S.; Chen, J.; Zuo, L.; Shi, M.; Zhan, X.; Li, C.-Z.; Chen, H. Nonfullerene Tandem Organic Solar Cells with High Open-Circuit Voltage of 1.97 V. *Adv. Mater.* **2016**, *28*, 9729–9734. [CrossRef] [PubMed]
58. Glass, A.M.; Von Der Linde, D.; Negran, T.J. High-voltage bulk photovoltaic effect and the photorefractive process in $LiNbO_3$. *Appl. Phys. Lett.* **1974**, *25*, 233–235. [CrossRef]
59. Koch, W.; Munser, R.; Ruppel, W.; Würfel, P. Bulk photovoltaic effect in $BaTiO_3$. *Solid State Commun.* **1975**, *17*, 847–850. [CrossRef]
60. Choi, T.; Lee, S.; Choi, Y.J.; Kiryukhin, V.; Cheong, S.-W. Switchable Ferroelectric Diode and Photovoltaic Effect in $BiFeO_3$. *Science* **2009**, *324*, 63–66. [CrossRef]
61. Yang, S.Y.; Seidel, J.; Byrnes, S.J.; Shafer, P.; Yang, C.-H.; Rossell, M.D.; Yu, P.; Chu, Y.-H.; Scott, J.F.; Ager, J.W.; et al. Above-bandgap voltages from ferroelectric photovoltaic devices. *Nat. Nanotechnol.* **2010**, *5*, 143–147. [CrossRef]
62. Horiuchi, S.; Tokura, Y. Organic ferroelectrics. *Nat. Mater.* **2008**, *7*, 357–366. [CrossRef]
63. Horiuchi, S.; Tokunaga, Y.; Giovannetti, G.; Picozzi, S.; Itoh, H.; Shimano, R.; Kumai, R.; Tokura, Y. Above-room-temperature ferroelectricity in a single-component molecular crystal. *Nat. Cell Biol.* **2010**, *463*, 789–792. [CrossRef]
64. Furukawa, T.; Nakajima, T.; Takahashi, Y. Factors governing ferroelectric switching characteristics of thin VDF/TrFE co-polymer films. *IEEE Trans. Dielectric. Electr. Insul.* **2006**, *13*, 1120–1131. [CrossRef]
65. Hartmann, W.J. Ferroelectric liquid crystal displays for television application. *Ferroelectrics* **1991**, *122*, 1–26. [CrossRef]
66. Sasabe, H.; Nakayama, T.; Kumazawa, K.; Miyata, S.; Fukada, E. Photovoltaic Effect in Poly(vinilydene fluoride). *Polym. J.* **1981**, *13*, 967–973. [CrossRef]
67. Sugita, A.; Suzuki, K.; Tasaka, S. Ferroelectric properties of a triphenylene derivative with polar functional groups in the crystalline state. *Phys. Rev. B* **2004**, *69*, 212201. [CrossRef]
68. Nakamura, M.; Horiuchi, S.; Kagawa, F.; Ogawa, N.; Kurumaji, T.; Tokura, Y.; Kawasaki, M. Shift current photovoltaic effect in a ferroelectric charge-transfer complex. *Nat. Commun.* **2017**, *8*, 1–6. [CrossRef]
69. Meyer, R.B.; Liebert, L.; Strzelecki, L.; Keller, P. Ferroelectric liquid crystals. *J. Phys. Lett.* **1975**, *36*, 69–71. [CrossRef]
70. Clark, N.A.; Lagerwall, S.T. Submicrosecond bistable electro-optic switching in liquid crystals. *Appl. Phys. Lett.* **1980**, *36*, 899–901. [CrossRef]
71. Bock, H.; Helfrich, W. Two ferroelectric phases of a columnar dibenzopyrene. *Liq. Cryst.* **1995**, *18*, 387–399. [CrossRef]
72. Kishikawa, K.; Nakahara, S.; Nishikawa, Y.; Kohmoto, S.; Yamamoto, M. A Ferroelectrically Switchable Columnar Liquid Crystal Phase with Achiral Molecules: Superstructures and Properties of Liquid Crystalline Ureas. *J. Am. Chem. Soc.* **2005**, *127*, 2565–2571. [CrossRef]
73. Miyajima, D.; Araoka, F.; Takezoe, H.; Kim, J.; Kato, K.; Takata, M.; Aida, T. Ferroelectric Columnar Liquid Crystal Featur-ing Confined Polar Groups within Core–Shell Architecture. *Science* **2012**, *336*, 209–213. [CrossRef]
74. Funatsu, Y.; Sonoda, A.; Funahashi, M. Ferroelectric liquid-crystalline semiconductors based on a phenylterthiophene skel-eton: Effect of introduction of oligosiloxane moieties and photovoltaic effect. *J. Mater. Chem. C* **2015**, *3*, 1982–1993. [CrossRef]
75. Seki, A.; Funahashi, M. Photovoltaic Effects in Ferroelectric Liquid Crystals based on Phenylterthiophene Derivatives. *Chem. Lett.* **2016**, *45*, 616–618. [CrossRef]
76. Seki, A.; Funatsu, Y.; Funahashi, M.; Seki, A. Anomalous photovoltaic effect based on molecular chirality: Influence of enantiomeric purity on the photocurrent response in π-conjugated ferroelectric liquid crystals. *Phys. Chem. Chem. Phys.* **2017**, *19*, 16446–16455. [CrossRef]

77. Seki, A.; Funahashi, M. Chiral photovoltaic effect in an ordered smectic phase of a phenylterthiophene derivative. *Org. Electron.* **2018**, *62*, 311–319. [CrossRef]
78. Funahashi, M.; Mori, Y. Linearly polarized electroluminescence device in which the polarized plane can be rotated electri-cally using a chiral liquid crystalline semiconductor. *Mater. Chem. Front.* **2020**, *4*, 2137–2148. [CrossRef]
79. Mori, Y.; Funahashi, M. Bulk photovoltaic effect in organic binary systems consisting of a ferroelectric liquid crystalline semiconductor and fullerene derivatives. *Org. Electron.* **2020**, *87*, 105962. [CrossRef]
80. Seki, A.; Yoshio, M.; Mori, Y.; Funahashi, M. Ferroelectric Liquid-Crystalline Binary Mixtures Based on Achiral and Chiral Trifluoromethylphenylterthiophenes. *ACS Appl. Mater. Interfaces* **2020**, *12*, 53029–53038. [CrossRef]
81. Anetai, H.; Wada, Y.; Takeda, T.; Hoshino, N.; Yamamoto, S.; Mitsuishi, M.; Takenobu, T.; Akutagawa, T. Fluorescent Ferro-electrics of Hydrogen-Bonded Pyrene Derivatives. *J. Phys. Chem. Lett.* **2015**, *6*, 1813–1818. [CrossRef] [PubMed]
82. Zhang, C.; Nakano, K.; Nakamura, M.; Araoka, F.; Tajima, K.; Miyajima, D. Noncentrosymmetric Columnar Liquid Crystals with the Bulk Photovoltaic Effect for Organic Photodetectors. *J. Am. Chem. Soc.* **2020**, *142*, 3326–3330. [CrossRef]

symmetry

MDPI

Article

Polar Alkoxy Group and Pyridyl Effects on the Mesomorphic Behavior of New Non-Symmetrical Schiff Base Liquid Crystals

Sayed Z. Mohammady [1,2,*], Daifallah M. Aldhayan [1], Mohammed A. Alshammri [1], Ayoub K. Alshammari [1], Mohammed Alazmi [3], Kanubhai D. Katariya [4], Mariusz Jaremko [5] and Mohamed Hagar [3,*]

[1] Department of Chemistry, College of Science, King Saud University, Riyadh 11451, Saudi Arabia; aldhayan@ksu.edu.sa (D.M.A.); Alshammri@ksu.edu.sa (M.A.A.); 442106682@student.ksu.edu.sa (A.K.A.)
[2] Department of Chemistry, Faculty of Science, Cairo University, Giza 12613, Egypt
[3] Chemistry Department, Faculty of Science, Alexandria University, Alexandria 21321, Egypt; Mohammadalazmmi@gmail.com
[4] Department of Chemistry, Faculty of Science, The Maharaja Sayajirao University of Baroda, Vadodara 390002, India; kanukatariya-chemphd@msubaroda.ac.in
[5] Biological and Environmental Sciences & Engineering Division (BESE), King Abdullah University of Science and Technology (KAUST), Thuwal 23955-6900, Saudi Arabia; mariusz.jaremko@kaust.edu.sa
* Correspondence: sahmed2.c@ksu.edu.sa (S.Z.M.); mohamed.hagar@alexu.edu.eg (M.H.)

Abstract: A series of non-symmetrical Schiff base liquid crystals were prepared and investigated. Schiff bases of p-alkyloxy aniline derivatives and 4-phenyl pyridine-4′-carbaldehyde were synthesized. The terminal alkoxy groups substituting aniline are of varied chain length, namely C6, C8, and C16. The structures of the compounds were confirmed via ^{1}H NMR and ^{13}C NMR spectroscopy. Different mesophases of the samples were thermally and optically characterized by differential thermal analysis (DSC) and polarized optical microscopy (POM). All samples revealed enantiotropic smectic B (SmB) and smectic A (SmA) mesophases. The results obtained were further correlated with the density functional theory (DFT) theoretical calculations. The results are compared to a series of compounds bearing biphenyl moiety in their mesogens. The thermal stabilities of the different mesophase reduced upon the increment of the alkoxy chain length. The temperature ranges of both the smectic mesophases of new compounds bearing the 4-phenyl pyridine moiety are generally expanded higher than the other series. In addition, the total mesophase range is greater in the new compounds when compared to their biphenyl analogues. The DFT results were investigated in terms of the molecular geometries and the frontier molecular orbitals as well as the charge distribution mapping to show and illustrate the difference in the mesomorphic properties.

Keywords: liquid crystal; schiff base; biphenyl; 4-phenyl pyridine; DFT

Citation: Mohammady, S.Z.; Aldhayan, D.M.; Alshammri, M.A.; Alshammari, A.K.; Alazmi, M.; Katariya, K.D.; Jaremko, M.; Hagar, M. Polar Alkoxy Group and Pyridyl Effects on the Mesomorphic Behavior of New Non-Symmetrical Schiff Base Liquid Crystals. *Symmetry* **2021**, *13*, 1832. https://doi.org/10.3390/sym13101832

Academic Editors: György Keglevich, Shoichi Ishihara and Sadahito Uto

Received: 16 August 2021
Accepted: 22 September 2021
Published: 1 October 2021

Publisher's Note: MDPI stays neutral with regard to jurisdictional claims in published maps and institutional affiliations.

1. Introduction

A unique state of matter called the Liquid crystalline state gained much importance and attracted considerable attention both in the basic sciences and technological applications because of a variety of advanced applications including digital displays, sensors, liquid crystal displays (LCDs), high-performance polymers, transporting of electron, ion or molecule and drug-delivery systems, and hybrid composites [1–5]. Rod-like liquid crystalline materials called calamitic mesogens (possessing a rigid core with two or more aromatic rings linked by connecting groups and flexible terminal chains), have attracted remarkable attention over recent decades due to their ability to display a wide range of technologically important liquid crystal phases [6–9]. With a variety of structures of rod-like molecules, they display some potential applications by changing their physical properties including dielectric anisotropy ($\Delta\varepsilon$), birefringence (Δn), rotational viscosity and mesophase range for liquid crystal compounds [10–13].

It was observed that the mesomorphic properties of thermotropic calamitic LC are mostly influenced by the nature of terminal chains including alkyl, alkyloxy, perfluorinated chain, ester or acyl group, or introduction of heteroatom into terminal chains [14–17]. Recently, a huge number of structurally varied thermotropic liquid crystals consist of polar heteroatoms/heterocyclic systems were designed and synthesized [17–19]. A variety of materials synthesized based on five and six membered heterocyclic rings comprising one, two, or three nitrogen atoms gave rise to attracting mesomorphic properties. The enhanced attractive forces and layer formation properties due to molecular stacking and packing of these liquid crystals are attributed to the position of nitrogen atoms and the availability of their lone pair of electrons. The inclusion of a heteroatom or heterocyclic unit induces substantial effects such as polarizability, enhanced lateral and/or longitudinal dipole moment, increased intermolecular interaction, and impact the mesophase stability, phase transition temperatures and introduction of transverse dipole moment, often stimulating negative dielectric anisotropic properties due to the existence of lone-pair electrons on heteroatoms of rings [20]. The mesomorphic compounds with five-membered heterocyclic units such as 1,3,4-oxadiazole [21–23], 1,2,4-oxadiazole [24,25], isoxazole [26,27], 1,3,4-thiadiazole [28,29], and thiophene [30–32] showed interesting mesomorphic properties. Recently, several structurally diverse liquid crystals based on pyridine were prepared and studied for their mesomorphic properties [33–35].

Furthermore, several liquid crystalline compounds with connecting units including imine, ester, ether, and azo to connect various aromatic rings were prepared and studied for their liquid crystal behavior [36–40]. Since the discovery of 4-methoxybenzylidene-4′-butylaniline at room temperature nematogen [41], much attention has been focused by the researchers as the introduction of Schiff base unit (-CH=N-) promote the stability of mesophase by maintaining the linearity and rigidity. Recently, several thermotropic mesogens based on Schiff base with low molar mass have been synthesized, characterized and investigated [32,42,43].

Mesomorphic behavior of rod-like liquid crystal compounds is essentially reliant on the shape of molecule depending on molecular conformation [44,45]. Even a minor alteration in the structure of mesogens results in changes in their mesomorphic properties significantly. Moreover, the formation of mesophase, type of phase and its thermal stability largely dependent on the central connecting unit and the terminal groups/chains attached to mesogenic compounds [44,46]. Consequently, in the design of new mesogenic compounds the linking unit, terminal groups and flexible hydrocarbon chains are crucial.

2. Materials and Methods

All chemicals were purchased from TCI Company, Japan. Their purity is higher than 98%. Schiff bases were prepared and recrystallized twice from ethanol/water mixture and were checked to be TLC pure, Scheme 1.

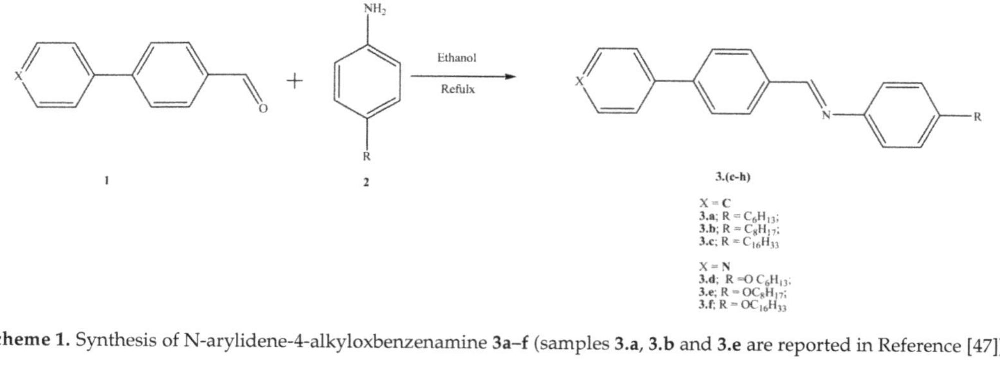

Scheme 1. Synthesis of N-arylidene-4-alkyloxbenzenamine **3a–f** (samples **3.a**, **3.b** and **3.e** are reported in Reference [47]).

Synthesis of N-Arylidene-4-alkyloxybenzenamine (**3a–f**)

Alkyloxybenzaldehyde (4.1 mmol) and 4-arylamine (4.1 mmol) equimolar equivalents were dissolved in ethanol (10 mL) and refluxed for two hours. The mixture was filtered after cooling to room temperature. TLC analysis showed that the obtained solid was washed with cold ethanol and recrystallized twice from hot ethanol to yield pure compounds.

(E)-*N*-(4-(Hexyloxy)phenyl)-1-(4-(pyridin-4-Yl)phenyl)methanimine (3.d)

Yield = 97%, IR (**KBr, cm^{-1}**): 3041, 2952, 2851 (C-H), 1625 (CH=N), 1597, 1539, 1511, 1408, 1361, 1270, 1183, 1118, 887, 822, 808, 721; **^{1}H NMR (400 MHz, $CDCl_3$, δ ppm)**: 0.93 (t, *J* = 7.2 Hz, 3H), 1.34–1.39 (m, 4H), 1.45–1.52 (m, 2H), 1.78–1.85 (m, 2H), 3.99 (t, *J* = 6.4 Hz, 2H), 6.96 (dd, *J* = 6.8 Hz, 2H), 7.29 (dd, *J* = 6.4 *Hz*, 2H), 7.57 (dd, *J* = 4.4 Hz, 2H), 7.77 (d, *J* = 8.0 Hz, 2H), 8.01 (d, *J* = 8.4 Hz, 2H), 8.55 (s, 1H), 8.71 (dd, *J* = 4.4 Hz, 2H); **^{13}CNMR (100 MHz, $CDCl_3$, δ ppm)**: 14.10, 22.65, 25.76, 29.28, 31.63, 68.29, 115, 121.57, 122.33, 127.35, 129.24, 137.12, 140.33, 144.30, 147.46, 150.42, 157.00, 158.17.

(E)-*N*-(4-(Octyloxy)phenyl)-1-(4-(pyridin-4-Yl)phenyl)methanimine (3.e)

Yield = 95%, **IR (KBr, cm^{-1})**: 3045, 2951, 2852 (C-H), 1621 (CH=N), 1594, 1533, 1505, 1402, 1361, 1275, 1181, 1117, 889, 822, 804, 721; **^{1}H NMR (400 MHz, $CDCl_3$, δ ppm)**: 0.91 (t, *J* = 6.8 Hz, 3H), 1.31–1.37 (m, 8H), 1.45–1.52 (m, 2H), 1.78–1.85 (m, 2H), 3.99 (t, *J* = 6.8 Hz, 2H), 6.96 (dd, *J* = 6.8 Hz, 2H), 7.29 (dd, *J* = 6.8 *Hz*, 2H), 7.57 (dd, *J* = 4.4 Hz, 2H), 7.75 (d, *J* = 8.0 Hz, 2H), 8.02 (d, *J* = 8.4 Hz, 2H), 8.55 (s, 1H), 8.71 (dd, *J* = 4.4 Hz, 2H); **^{13}CNMR (100 MHz, $CDCl_3$, δ ppm)**:14.16, 22.70, 26.08, 29.28, 29.32, 29.41, 31.85, 68.30, 115.00, 121.57, 122.33, 127.35, 129.24, 137.13, 140.34, 144.31, 147.46, 150.42, 157.00, 158.17.

(E)-*N*-(4-(Hexadecyloxy)phenyl)-1-(4-(pyridin-4-Yl)phenyl)methanimine (3.f)

Yield = 91%, **IR (KBr, cm^{-1})**: 3038, 2955, 2847 (C-H), 1618 (CH=N), 1595, 1534, 1501, 1405, 1365, 1277, 1182, 1115, 887, 822, 804, 720; **^{1}H NMR (400 MHz, $CDCl_3$, δ ppm)**: 0.90 (t, *J* = 6.4 Hz, 3H), 1.28 (m, 26H), 1.73–1.85 (m, 2H), 4.00 (t, *J* = 6.4 Hz, 2H), 6.96 (dd, *J* = 8.4 Hz, 2H), 7.29 (d, *J* = 8.4 *Hz*, 2H), 7.57 (d, *J* = 5.6 Hz, 2H), 7.77 (d, *J* = 8.4 Hz, 2H), 8.02 (d, *J* = 8.0 Hz, 2H), 8.56 (s, 1H), 8.71 (d, *J* = 5.6 Hz, 2H).

^{1}H NMR and ^{13}C NMR results (Figures S1–S5) are given in the Supplementary Materials as following: Figure S1: ^{1}H NMR of compound (OC6) **3.d**; Figure S2: ^{13}C NMR of compound (OC6) **3.d**; Figure S3: ^{1}H NMR of compound (OC8) **3.e**; Figure S4: ^{13}C NMR of compound (OC8) **3.e**; Figure S5: ^{1}H NMR of compound **3a** (OC16) **3.f**.

The phase changes in the materials were determined by means of differential scanning calorimetry (DSC), DSC-60A, Shimadzu, Japan. The size of samples was in the range of 2–3 mg. Samples were encapsulated in Al pans and were heated or cooled in inert atmosphere (dry nitrogen atmosphere). Measurements were accomplished at 10.0 °C/min. Samples were heated from room temperature to 200 °C and cooled back to room temperature at the same heating. The transition temperatures for the various phase changes were gained from the endothermic peak minima of enthalpy. The accuracy of temperature recording was better than 1.0 °C.

3. Results and Discussion

3.1. Synthesis and Characterization

The ^{1}H NMR of compounds showed that the terminal methyl group protons were observed as a triplet at δ 0.91 ppm whereas the protons of methylene group directly attached to the aromatic ring observed at δ 2.99 ppm as a triplet. The appearance of singlet at δ 8.55 ppm corresponding to imine linkage proton (-CH=N-) confirmed the formation of Schiff base for prepared compounds. The protons near the nitrogen of the pyridine ring were observed as a doublet most upfield at δ 8.71 ppm. The rest of the aromatic and aliphatic protons were observed as per the substitution pattern of compounds. In the ^{13}C NMR spectra of new compounds, the terminal methyl carbon was observed most upfield δ 14.10 ppm while methylene carbon directly attached to oxygen were observed at δ 68.29 ppm. The carbon of the imine linkage was appeared most downfield at δ 158.17 ppm whereas two carbons in vicinity to nitrogen of pyridine ring were observed at δ 150.42 ppm.

The rest of the carbons were observed as per the substitution pattern and in agreement with the structures of the new compounds.

3.2. Mesomorphic Behavior

Transition temperatures and the corresponding associated enthalpy changes of the prepared compounds were measured by DSC and displayed in Table 1. The DSC thermograms of the three new alkoxy derivative (**3.d**, **3.e** and **3.f**) are presented in Figure 1a–c. All the prepared compounds exhibited dimorphic enantiotropic mesomorphic characteristics whereby the corresponding endotherms were regarded as following according to the increase in temperature: crystalline to smectic B (Cryst.-SmB), SmB to smectic A (SmB-SmA) and SmA-isotropic (SmA-I) transitions. The SmB and SmA mesophases were specified by the existence of the standard Schlieren and focal-conic fan textures.

Table 1. Phase transitions: temperatures (T, °C) and enthalpies (ΔH, kJ/mol) for the prepared compounds **3.a–f**. The abbreviations Cr-SmB, SmB-SmA, SmB-I, SmA-I, SmA-N, and N-I refer to the crystalline to smectic B, smectic B smectic A, smectic B to isotropic, smectic A to isotropic, smectic A to nematic and nematic to isotropic transitions, respectively. ΔH refers to the enthalpy changes in kJ/mol for the different phase transitions.

Compound	°C	kJ/mol	°C	kJ/mol	°C	kJ/mol	°C	kJ/mol	°C	kJ/mol
	$T_{Cr\text{-}SmB}$	$\Delta H_{Cr\text{-}SmB}$	$T_{SmB\text{-}SmA}$	$\Delta H_{SmB\text{-}SmA}$	$T_{SmB\text{-}I}$	$\Delta H_{SmA\text{-}I}$	$T_{SmA\text{-}N}$	$\Delta H_{SmA\text{-}N}$	$T_{N\text{-}I}$	$\Delta H_{N\text{-}I}$
3.a (**C6**)	86.2	28.09					128.0	—	135.0	—
3.b (**C8**)	95.1	76.02	130.0	3.03			132.2	1.77	134.5	0.46
3.c (**C16**)	110.9	119.75	112.7	3.78	118.6	3.71				
3.d (**-OC$_6$**)	121.8	35.78	149.9	3.23	170.1	5.61				
3.e (**-OC$_8$**)	113.9	21.23	146.0	2.37	168.1	3.76				
3.f (**-OC$_{16}$**)	112.3	24.62	143.3	2.13	166.6	3.42				

It is worth noting that this study presents a continuation of our systematic work in which we investigate the effect of conjugation length on the mesomorphic behavior on model examples of LCs. In a recent article, we have investigated the effect of introducing 4-phenyl pyridine moiety on LC molecules terminated with alkyl group [48]. The mesophases of new compounds were verified by observing their textures through POM (Figure 2). Observation of mesomorphic behavior of the new compounds bearing the 4-phenylpyridyl mesogens (**3.d–f**) under the POM showed enantiotropic dimorphic mesomorphism. After heating from crystalline phase to isotropic liquid and again cooling it from its isotropic liquid phase to crystalline phase, all new compounds verify the presence of two mesophases both via POM and DSC thermograms (Table 1). First, SmB phase texture (Figure 2a appears after melting, and later it turned out to SmA phase (Figure 2b) at elevated temperatures. The SmB and SmA mesophases were identified by the appearance of the standard Schlieren and focal-conic fan textures. All observed LC mesophase textures agree with the literature both with respect to their textures and the associated enthalpy changes recorded by DSC measurements (ΔH values lies between 2.13 and 5.61 kJ/mol).

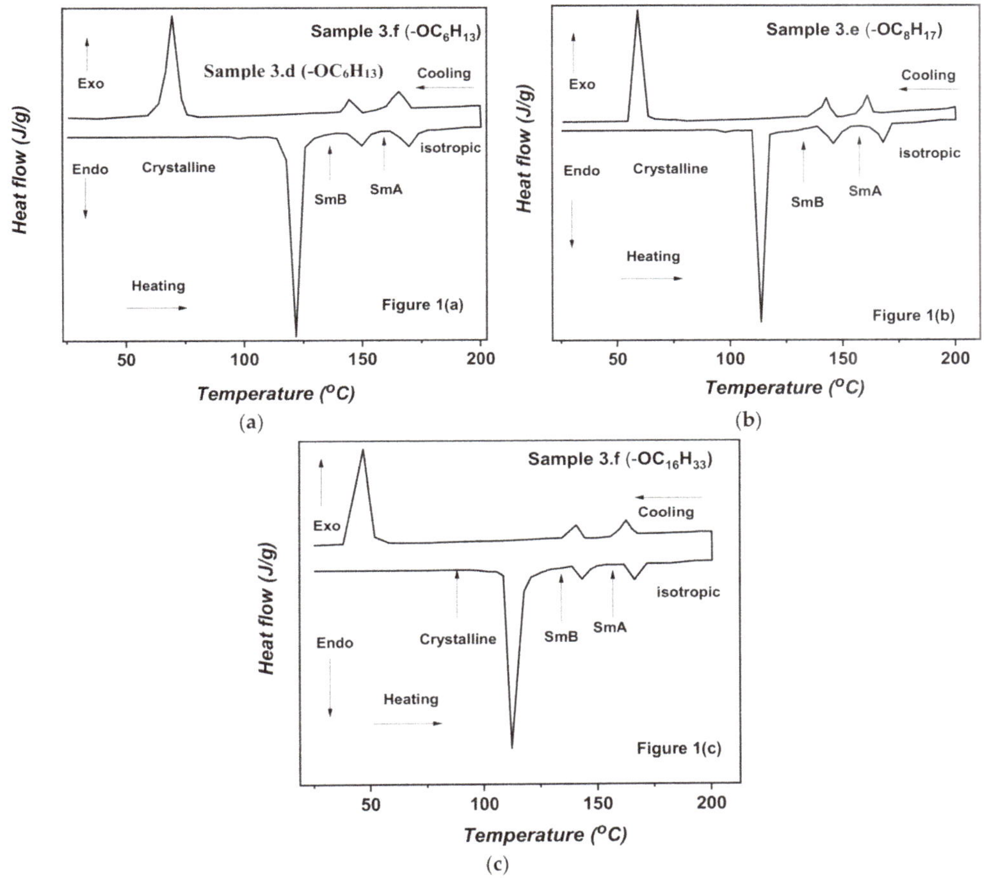

(a) (b) (c)

Figure 1. (**a–c**) DSC thermograms of the new compounds **3.d**, **3.e** and **3.f** during heating and cooling cycles (10 °C min^{-1}).

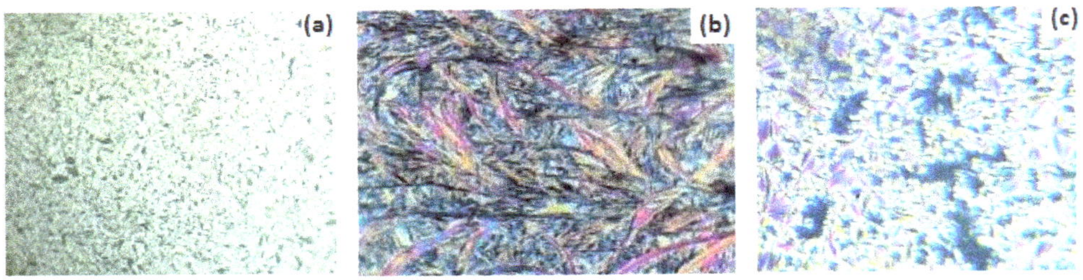

Figure 2. Mesophase textures observed by POM during heating cycle of compound **3.d** Smectic B phase at 130.0 °C (**a**), **3.e** Smectic A phase at 158.0 °C (**b**), **3.f** Smectic A phase at 162.0 °C (**c**).

The effect of variation of the alkyloxy chain length on the mesomorphic behavior of the new compounds is displayed in Figure 3.

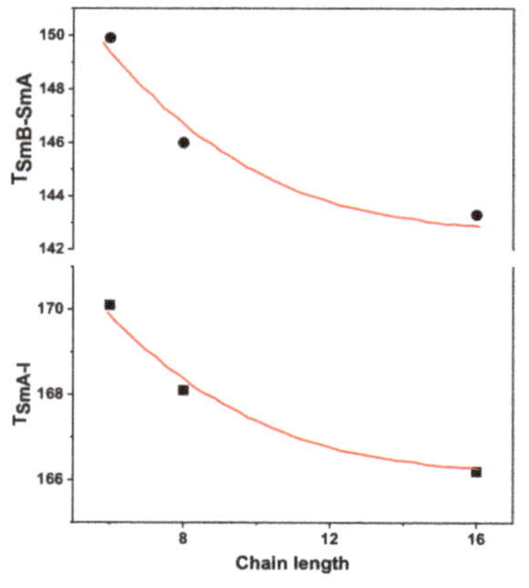

Figure 3. Dependence of the mesophase transition temperatures on the alkyloxy chain length.

It is observed that the transition temperatures of SmB/SmA mesophase transition decreases as alkoxy chain length increase. In the same manner, the SmA-I transition temperatures showed similar response upon the increase of the terminal alkoxy group (C6 to C16).

The variation of mesophase temperature ranges of the different observed textures (SmB and SmA) in the new compounds **3.d** to **3.f** is compared to the results obtained for mesomprohic compounds bearing biphenyl moiety and terminated with long chain alkyl groups rather than our new series containing 4-phenyl pyridyl group in their mesogens and terminated with long chains of alkoxy groups. The results are shown in Table 2.

Table 2. Variation of mesophase temperature ranges of the different observed textures for the mesomorphic compounds bearing biphenyl group (alkyl terminated) rather than 4-phenyl pyridine (alkoxy terminated) ends in their mesogens.

Alkyl Chain Length	SmB Range (°C)		SmA Range (°C)	
	Biphenyl	4-Phenyl Pyridine	Biphenyl	4-Phenyl Pyridine
C6	41.8	28.1	–	20.2
C8	34.8	35.1	2.2	22.1
C16	1.8	31.0	5.9	23.3

The increased mesophase stabilities and broadening in the SmA mesomorphic temperature range in the compounds **3d–f** can be rationalized to two different factors. The first factor is the increased pi-pi stacking arising from the extra contribution of the N-atom of the pyridine ring to the total conjugated electrons along the mesogen. The second factor is the presence of two lone pairs of electrons on the oxygen atom of the alkoxy chains which is directly fused to the mesogenic group. The presence of the lone pairs of electrons on the oxygen atom may lead to extra conjugation which in turn increases the length of the mesogenic group (increased pi-pi stacking). Consequently, the mesophase stability is largely increased in alkoxy terminated series of LCs compared to the biphenyl series ended with alkyl chains. This elevation in pi-pi stacking can explain the promoted SmB and SmA mesomorphic order rather than the nematic one even for the sample bearing relatively short hexyloxy chain length (**3.d**). Comparing the effect in the present study with LC having 4-phenyl pyridine and terminated with alkyl groups, the effects of introducing the above explained two factors were more pronounced [48].

3.3. DFT Calculations

Gaussian 9 was used to carry out theoretical DFT computations in the gas phase using DFT and B3LYP 6-311g (d,p) base site. All optimized structures were found to be stable due to the lack of the imaginary frequency. The results of the theoretical calculations of the studied compounds **3.a–f** reveal a non-planar geometry (Figure 4). The phenyl rings, on the other hand, are not coplanar, and the twist angle of the CH=N is affected presence of the pyridine atom that affect the degree of aromaticity conjugated arylidene component. Furthermore, the biphenyl ring and its pyridyl derivative are significantly different in the degree of the aromaticity and consequently the presence of N- heteroatom affects the angle deviated from co-planarity. The twist CH=N angle (θ) = 7.8° for the hexyl derivative **3.a** of the biphenyl and 4.6° for its pyridyl derivative **3.d**. The pyridyl derivative **3.e** (θ) = 7.8° had the higher twist angle, while its longest chain length (C=16) derivative **3.f** had twist angle (θ) = 9.3°. The result of the difference in twist angle of the alkoxy derivatives of the pyridyl and its alkyl derivatives of the biphenyl one could be explained in terms of better conjugation of the π-cloud of aromatic rings with the imino group with the alkoxy one of higher mesomeric resonance effect rather than the hyper conjugated alkyl group, and as a result of the high conjugation, arylidene system is more enforced to be more planar in the case of the alkoxypyridyl derivatives. While these theoretical molecular geometries may be a good predictor of the preferred molecular structure in the gas phase, the presence of these compounds in liquid crystalline matter condensed phases with a different lowest energy and more elongated species will be favored [49].

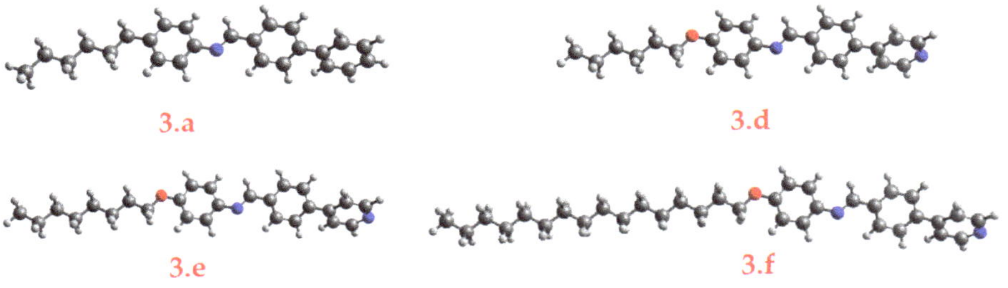

Figure 4. Optimized geometrical structure of the prepared compounds **3.a,d–f**.

3.4. Frontier Molecular Orbitals (FMOs)

The results of the calculated plots of the synthesized compounds frontier molecular orbitals HOMO (highest occupied) and LUMO (lowest unoccupied) are shown in Table 3 and Figure 5. Figure 5 shows how the electron densities of the sites involved in LUMO and HOMO generation are localized on the pyridyl part and its conjugated rings. On the other hand, the electron densities of the FMOs were affected by the degree of conjugated rings affected by the polar alkoxy group as well as the presence of the hetaryl atom. Furthermore, the degree of conjugation and the existence of the N-atom had a substantial impact on the energy levels of the FMOs and the energy gap between them. The presence of a more aromatic ring in the biphenyl derivatives affects the levels of the frontier molecular orbitals. The energy level of the FMOs is dramatically reduced when the N-atom replaces the carbon. Furthermore, the energy difference between the FMOs was significantly altered by the extraconjugation of the alkoxy derivatives **3.d–f** due to the resonance effect. The pyridyl derivative, on the other hand, showed a greater decrease in the energy gap than the phenyl derivative. This was demonstrated by increases in co-planarity in the case of the pyridyl derivative with respect to its phenyl, which resulted in additional more conjugation and a decrease in the FMOs energy gab. On the other hand, there is no difference in the FMOs energy gab and energy difference with the longer chain lengths.

Table 3. FMO Energies e. V. and their values for the prepared compounds **3.a,d–f**.

Compounds	HOMO	LUMO	ΔE
3.a	−5.75	−1.85	3.90
3.d	−5.77	−2.10	3.67
3.e	−5.76	−2.10	3.66
3.f	−5.77	−2.11	3.66

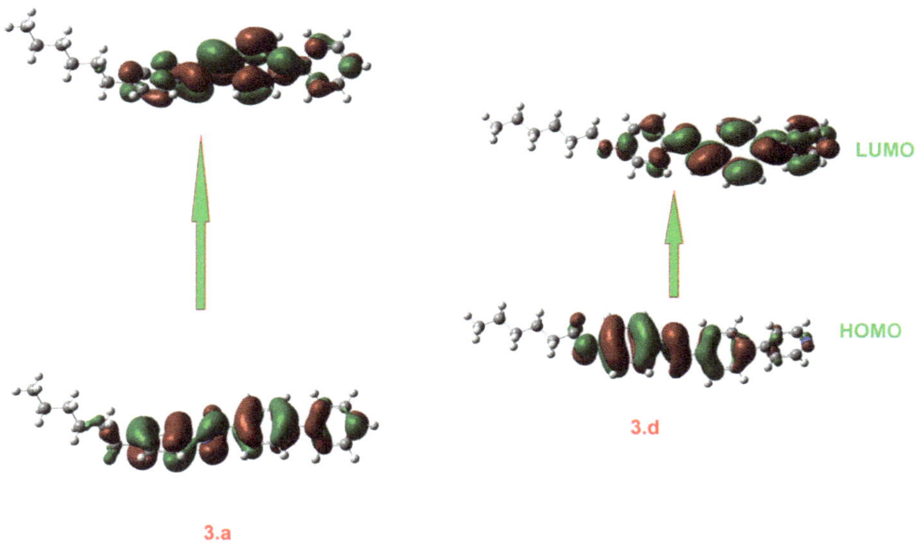

Figure 5. The calculated plots for the prepared compounds **3.a,d** of the frontier molecular orbitals.

3.5. Molecular Electrostatic Potential (MEP)

According to molecular electrostatic potential (MEP), the charge distribution map for the synthesized compounds **3.a,d–f** was estimated using the same method of calculations with the same basis sets as shown in Figure 6. Furthermore, the molecular electrostatic potential (MEP) is an important parameter to understand when analyzing the electron density distribution over molecules [49–51]. MEP is also one of the one of excellent tool to predict the intermolecular and/or intramolecular interactions. The existence of the polar N-atom and the connecting CH=N group of the examined compounds had an impact on the localization of the iso-electronic density of the electron rich and electron deficient areas. The Schiff base part CH=N group has the lowest electron density, whereas the terminal alkoxy chain has the maximum electron density. The anticipated charge distribution maps could be used to illustrate how liquid crystals molecular structure assist mesophase behavior change.

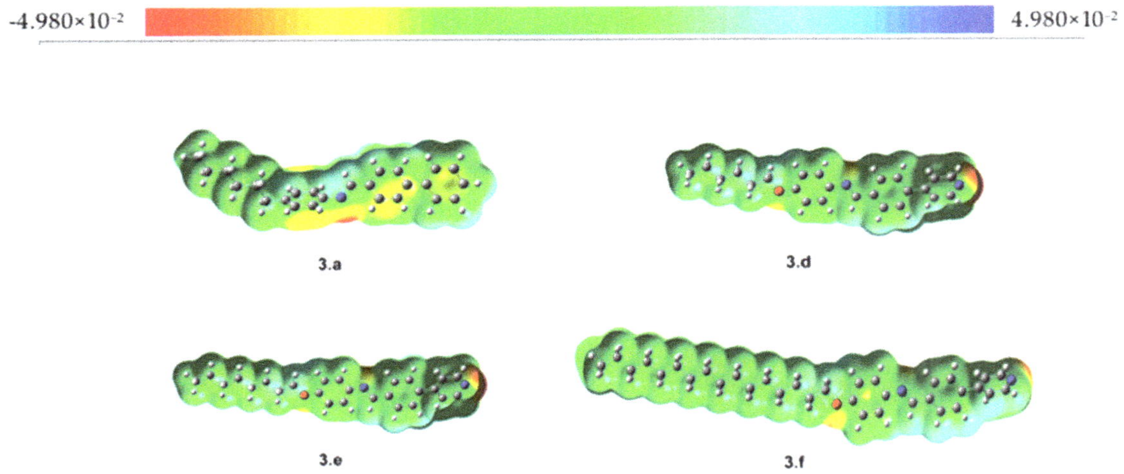

Figure 6. Molecular electrostatic potentials (Hartree) for the prepared compounds **3.a**,**d**–**f**.

4. Conclusions

A novel series of Schiff base liquid crystals were prepared and investigated. Schiff bases of p-alkyloxy aniline derivatives and 4-phenylpyridine-4′-carbaldehyde were synthesized. The terminal alkoxy groups substituting aniline are of varied chain length, namely C6, C8, and C16. All samples revealed enantiotropic smectic B (SmB) and smectic A (SmA) mesophases. The results are compared to two series of compounds bearing pyridine and biphenyl, respectively, in their mesogens. The thermal stabilities of the different mesophase reduced upon the increment of the alkoxy chain length. The temperature ranges of both the smectic mesophases of new compounds bearing the 4-phenyl pyridine moiety are generally expanded higher than the other two series. In addition, the total mesophase range is greater in the new compounds when compared to their biphenyl analogues. The increased mesophase stabilities and broadening in the SmA mesomorphic temperature range in the new compounds were rationalized to the increased pi-pi stacking arising from the extra contribution of the N-atom of the pyridine ring. A second factor that may contribute effectively is the lone pairs of electrons on the oxygen atom of the alkoxy chains. The presence of the lone pairs of electrons on the oxygen atom imparted extra conjugation and consequently, the mesophase stability increased. The results obtained were in a good agreement with the density functional theory (DFT) theoretical calculations. The result of DFT showed a difference in twist angle as well as the charge distribution maps of the alkoxy derivatives of the pyridyl and its alkyl derivatives and this could explain the difference in the mesomorphic properties.

Supplementary Materials: ^{1}H NMR and ^{13}C NMR results of the samples **3.d**, **3.e** and **3.f** are given as supplementary materials. They are available at https://www.mdpi.com/article/10.3390/sym13101832/s1, Figure S1: ^{1}H NMR of compound (OC6) **3.d**; Figure S2: ^{13}C NMR of compound (OC6) **3.d**; Figure S3: ^{1}H NMR of compound (OC8) **3.e**; Figure S4: ^{13}C NMR of compound (OC8) **3.e**; Figure S5: ^{1}H NMR of compound **3a** (OC16) **3.f**.

Author Contributions: Conceptualization, S.Z.M. and M.H.; methodology, S.Z.M., D.M.A., M.A.A., A.K.A., M.A., K.D.K. and M.H.; validation, S.Z.M. and M.H.; investigation, S.Z.M., D.M.A., K.D.K., M.J. and M.H.; resources, S.Z.M. and M.H.; data curation, S.Z.M.; writing—original draft preparation, S.Z.M. and M.H.; writing—review and editing, S.Z.M., M.J. and M.H.; visualization, S.Z.M.; supervision, S.Z.M. and M.H.; project administration, S.Z.M.; funding acquisition, S.Z.M. All authors have read and agreed to the published version of the manuscript.

Funding: This research was funded by the Deanship of Scientific Research at King Saud University for funding this work through the research group number (RG-1440-143).

Acknowledgments: The authors extend their appreciation to the Deanship of Scientific Research at King Saud University for funding this work through the research group number (RG-1440-143).

Conflicts of Interest: The authors declare no conflict of interest.

References

1. Alberti, G.; Atwood, J.L. *Comprehensive Supramolecular Chemistry. 7. Solid-State Supramolecular Chemistry: Two-And Three-Dimensional Inorganic Networks*; Pergamon Press: Oxford, UK, 1996.
2. Schenning, A.; Crawford, G.P.; Broer, D.J. *Liquid Crystal Sensors*; CRC Press: Boca Raton, FL, USA, 2017.
3. Picken, S.; Sikkema, D.; Boerstoel, H.; Dingemans, T.; Van der Zwaag, S. Liquid crystal main-chain polymers for high-performance fibre applications. *Liq. Cryst.* **2011**, *38*, 1591–1605. [CrossRef]
4. Müller-Goymann, C. Physicochemical characterization of colloidal drug delivery systems such as reverse micelles, vesicles, liquid crystals and nanoparticles for topical administration. *Eur. J. Pharm. Biopharm.* **2004**, *58*, 343–356. [CrossRef]
5. Spengler, M.; Dong, R.Y.; Michal, C.A.; Hamad, W.Y.; MacLachlan, M.J.; Giese, M. Hydrogen-Bonded Liquid Crystals in Confined Spaces—Toward Photonic Hybrid Materials. *Adv. Funct. Mater.* **2018**, *28*, 1800207. [CrossRef]
6. De Gennes, P.-G.; Prost, J. *The Physics of Liquid Crystals*; Oxford University Press: Oxford, UK, 1993.
7. Hird, M. Fluorinated liquid crystals–properties and applications. *Chem. Soc. Rev.* **2007**, *36*, 2070–2095. [CrossRef]
8. Chigrinov, V.G. *Liquid Crystal Devices: Physics and Applications*; Artech House Publishers: Norwood, MA, USA, 1999.
9. Patil, N.V.; Sahoo, R.; Veerabhadraswamy, B.; Chakraborty, S.; Dhar, R.; Mathad, R.; Yelamaggad, C. Polarization of three-ring Schiff base ferroelectric liquid crystals. *Liq. Cryst.* **2020**, *48*, 1194–1205. [CrossRef]
10. Si, G.; Zhao, Y.; Leong, E.S.P.; Liu, Y.J. Liquid-crystal-enabled active plasmonics: A review. *Materials* **2014**, *7*, 1296–1317. [CrossRef] [PubMed]
11. Wang, L.; Lin, X.-W.; Liang, X.; Wu, J.-B.; Hu, W.; Zheng, Z.-G.; Jin, B.-B.; Qin, Y.-Q.; Lu, Y.-Q. Large birefringence liquid crystal material in terahertz range. *Opt. Mater. Express* **2012**, *2*, 1314–1319. [CrossRef]
12. Yun, H.J.; Jo, M.H.; Jang, I.W.; Lee, S.H.; Ahn, S.H.; Hur, H.J. Achieving high light efficiency and fast response time in fringe field switching mode using a liquid crystal with negative dielectric anisotropy. *Liq. Cryst.* **2012**, *39*, 1141–1148. [CrossRef]
13. Lapanik, V.; Sasnouski, G.; Timofeev, S.; Shepeleva, E.; Evtyushkin, G.; Haase, W. New highly anisotropic liquid crystal materials for high-frequency applications. *Liq. Cryst.* **2018**, *45*, 1242–1249. [CrossRef]
14. Luo, C.-C.; Jia, Y.-G.; Sun, B.-F.; Meng, F.-B. The effect of chain length in terminal groups on the mesomorphic behavior of novel (−)-menthol-based chiral liquid crystal compounds with a blue phase. *New J. Chem.* **2017**, *41*, 3677–3686. [CrossRef]
15. Cheng, Z.; Li, K.; Wang, F.; Wu, X.; Chen, X.; Xiao, J.; Zhang, H.; Cao, H.; Yang, H. Influence of linkage and terminal group on the liquid crystalline and helical twisting behaviours of cholesteryl esters. *Liq. Cryst.* **2011**, *38*, 803–812. [CrossRef]
16. Hagar, M.; Ahmed, H.; Aouad, M. Mesomorphic and DFT diversity of Schiff base derivatives bearing protruded methoxy groups. *Liq. Cryst.* **2020**, *47*, 2222–2233. [CrossRef]
17. Hagar, M.; Ahmed, H.; Saad, G. New calamitic thermotropic liquid crystals of 2-hydroxypyridine ester mesogenic core: Mesophase behaviour and DFT calculations. *Liq. Cryst.* **2020**, *47*, 114–124. [CrossRef]
18. Regis, E.; de Aguiar, L.O.; Tuzimoto, P.; Girotto, E.; Frizon, T.E.; Dal Bó, A.G.; Zapp, E.; Marra, R.; Gallardo, H.; Vieira, A.A. Effect of heteroatom exchange (S/Se) in the mesomorphism and physical properties of benzochalcogenodiazole-based liquid crystals. *Dye. Pigment.* **2018**, *157*, 109–116. [CrossRef]
19. Ong, L.-K.; Ha, S.-T.; Yeap, G.-Y.; Lin, H.-C. Heterocyclic pyridine-based liquid crystals: Synthesis and mesomorphic properties. *Liq. Cryst.* **2018**, *45*, 1574–1584. [CrossRef]
20. Rodrigues, L.D.; Sunil, D.; Chaithra, D.; Bhagavath, P. 1, 2, 3/1, 2, 4-Triazole containing liquid crystalline materials: An up-to-date review of their synthetic design and mesomorphic behavior. *J. Mol. Liq.* **2020**, *297*, 111909. [CrossRef]
21. Saha, S.K.; Paul, M.K. Mesomorphic and photophysical behaviour of 1, 3, 4-oxadiazole based hockey stick reactive mesogens. *Liq. Cryst.* **2019**, *46*, 386–396. [CrossRef]
22. Santos, A.B.; Manfredi, A.M.; Salla, C.A.; Farias, G.; Girotto, E.; Eccher, J.; Westphal, E.; Curcio, S.F.; Cazati, T.; Malvestiti, I. Highly luminescent liquid crystals by connecting 1, 3, 4-oxadiazole with thiazolo [5, 4-d] thiazole units. *J. Mol. Liq.* **2021**, *321*, 114887. [CrossRef]
23. Saha, S.K.; Mohiuddin, G.; Paul, M.K.; Gupta, S.P.; Khan, R.K.; Ghosh, S.; Pal, S.K. Polar switching and cybotactic nematic ordering in 1, 3, 4-thiadiazole-based short-core hockey stick-shaped fluorescent liquid crystals. *ACS Omega* **2019**, *4*, 7711–7722. [CrossRef]
24. Fouad, F.S.; Ness, T.; Wang, K.; Ruth, C.E.; Britton, S.; Twieg, R.J. Biphenylyl-1, 2, 4-oxadiazole based liquid crystals–synthesis, mesomorphism, effect of lateral monofluorination. *Liq. Cryst.* **2019**, *46*, 2281–2290. [CrossRef]
25. Ali, G.Q.; Tomi, I.H.R. Synthesis and characterization of new mesogenic esters derived from 1, 2, 4-oxadiazole and study the effect of alkoxy chain length in their liquid crystalline properties. *Liq. Cryst.* **2018**, *45*, 421–430. [CrossRef]
26. Kotian, S.Y.; Mohan, C.D.; Merlo, A.A.; Rangappa, S.; Nayak, S.C.; Rai, K.L.; Rangappa, K.S. Small molecule based five-membered heterocycles: A view of liquid crystalline properties beyond the biological applications. *J. Mol. Liq.* **2020**, *297*, 111686. [CrossRef]

27. Fritsch, L.; Baptista, L.A.; Bechtold, I.H.; Araujo, G.; Mandle, R.J.; Merlo, A.A. Isoxazoline-and isoxazole-liquid crystalline schiff bases: A puzzling game dictated by entropy and enthalpy effects. *J. Mol. Liq.* **2020**, *298*, 111750. [CrossRef]
28. Olate, F.A.; Parra, M.L.; Vergara, J.M.; Barberá, J.; Dahrouch, M. Star-shaped molecules as functional materials based on 1, 3, 5-benzenetriesters with pendant 1, 3, 4-thiadiazole groups: Liquid crystals, optical, solvatofluorochromic and electrochemical properties. *Liq. Cryst.* **2017**, *44*, 1173–1184. [CrossRef]
29. Sharma, V.S.; Shah, A.P.; Sharma, A.S. A new class of supramolecular liquid crystals derived from azo calix [4] arene functionalized 1, 3, 4-thiadiazole derivatives. *New J. Chem.* **2019**, *43*, 3556–3564. [CrossRef]
30. Wan, D.; Shi, F.; Che, Z.; Mo, L.; Hu, M.; Guo, Q.; Li, J.; An, Z.; Li, J. Introduction of 5, 6-dihydro-4 H-cyclopenta [b] thiophene core unit into phenyl-tolane: Expanding the mesophase range and increasing the birefringence and dielectric anisotropy of liquid crystal. *Liq. Cryst.* **2021**. [CrossRef]
31. Wan, D.; Yang, X.; Li, J.; Hu, M.; Mo, L.; Che, Z.; Guo, Q.; An, Z.; Li, J. Synthesis and properties of 5, 6-dihydro-4H-cyclopenta [b] thiophene-based nematic liquid crystals: A new access to mesogens with high birefringence and large dielectric anisotropy. *J. Mol. Liq.* **2021**, *327*, 114827. [CrossRef]
32. Nafee, S.S.; Ahmed, H.; Hagar, M. Theoretical, experimental and optical study of new thiophene-based liquid crystals and their positional isomers. *Liq. Cryst.* **2020**, *47*, 1291–1302. [CrossRef]
33. Devadiga, D.; Ahipa, T. An Up-to-date Review on Halogen-Bonded Liquid Crystals. *J. Mol. Liq.* **2021**, *333*, 115961. [CrossRef]
34. Devadiga, D.; Ahipa, T. Heterodimeric hydrogen-bonded mesogens comprising pyridine moiety: A review. *Liq. Cryst. Rev.* **2020**, *8*, 5–28. [CrossRef]
35. Devadiga, D.; Ahipa, T. Recent synthetic advances in pyridine-based thermotropic mesogens. *RSC Adv.* **2019**, *9*, 23161–23228. [CrossRef]
36. Nakum, K.J.; Katariya, K.D.; Jadeja, R.N. Synthesis, characterization, and mesomorphic properties of some new Schiff base homologues series and their Cu (II) complexes. *Mol. Cryst. Liq. Cryst.* **2020**, *708*, 1–13. [CrossRef]
37. Nakum, K.J.; Katariya, K.D.; Jadeja, R.; Prajapati, A. Schiff base of 4-n-alkoxy-2-hydroxy benzaldehyde with 4-amino acetophenone and their Cu (II) complexes: Synthesis, characterization and mesomorphic behavior. *Mol. Cryst. Liq. Cryst.* **2019**, *690*, 1–13. [CrossRef]
38. Al-Hamdani, U.J.; Abbo, H.S.; Al-Jaber, A.A.; Titinchi, S.J. New azo-benzothiazole based liquid crystals: Synthesis and study of the effect of lateral substituents on their liquid crystalline behaviour. *Liq. Cryst.* **2020**, *47*, 2257–2267. [CrossRef]
39. Nafee, S.S.; Hagar, M.; Ahmed, H.A.; Alhaddad, O.; El-Shishtawy, R.M.; Raffah, B.M. New two rings Schiff base liquid crystals; ball mill synthesis, mesomorphic, Hammett and DFT studies. *J. Mol. Liq.* **2020**, *299*, 112161. [CrossRef]
40. Arakawa, Y.; Ishida, Y.; Tsuji, H. Ether-and Thioether-Linked Naphthalene-Based Liquid-Crystal Dimers: Influence of Chalcogen Linkage and Mesogenic-Arm Symmetry on the Incidence and Stability of the Twist–Bend Nematic Phase. *Chem.-A Eur. J.* **2020**, *26*, 3767–3775. [CrossRef]
41. Kelker, H.; Scheurle, B. A liquid-crystalline (nematic) phase with a particularly low solidification point. *Angew. Chem. Int. Ed. Engl.* **1969**, *8*, 884–885. [CrossRef]
42. Ahmed, H.; Mansour, E.; Hagar, M. Mesomorphic study and DFT simulation of calamitic Schiff base liquid crystals with electronically different terminal groups and their binary mixtures. *Liq. Cryst.* **2020**, *47*, 2292–2304. [CrossRef]
43. Nafee, S.S.; Hagar, M.; Ahmed, H.A.; El-Shishtawy, R.M.; Raffah, B.M. The synthesis of new thermal stable schiff base/ester liquid crystals: A computational, mesomorphic, and optical study. *Molecules* **2019**, *24*, 3032. [CrossRef]
44. Hagar, M.; Ahmed, H.; El-Sayed, T.; Alnoman, R. Mesophase behavior and DFT conformational analysis of new symmetrical diester chalcone liquid crystals. *J. Mol. Liq.* **2019**, *285*, 96–105. [CrossRef]
45. Hagar, M.; Ahmed, H.; Alhaddad, O. New azobenzene-based natural fatty acid liquid crystals with low melting point: Synthesis, DFT calculations and binary mixtures. *Liq. Cryst.* **2019**, *46*, 2223–2234. [CrossRef]
46. Ahmed, H.A.; Hagar, M.; El-Sayed, T.H.; Alnoman, R.B. Schiff base/ester liquid crystals with different lateral substituents: Mesophase behaviour and DFT calculations. *Liq. Cryst.* **2019**, *46*, 1–11. [CrossRef]
47. Srinivasu, C.; Pisipati, V.G.K.M.; Prabhu, C.R.; Murty, P.N.; Lakshiminarayana, S. Phase Transitions in p-(Phenyl Benzylidene)-p1-Alkylaniline Compounds: A Dilatometric Study. *Z. Naturforsch.* **2007**, *62*, 75–83. [CrossRef]
48. Zakaria, M.A.; Alazmi, M.; Katariya, K.D.; El Kilany, Y.; El Ashry, E.H.; Jaremko, M.; Hagar, M.; Mohammady, S.Z. Mesomorphic Behaviour and DFT Insight of Arylidene Schiff Base Liquid Crystals and Their Pyridine Impact Investigation. *Crystals* **2021**, *11*, 978. [CrossRef]
49. Politzer, P.; Murray, J.S. Relationships between dissociation energies and electrostatic potentials of CNO_2 bonds: Applications to impact sensitivities. *J. Mol. Struct.* **1996**, *376*, 419–424. [CrossRef]
50. Scrocco, E.; Tomasi, J. Electronic molecular structure, reactivity and intermolecular forces: An euristic interpretation by means of electrostatic molecular potentials. In *Advances in Quantum Chemistry*; Elsevier: Amsterdam, The Netherlands, 1978; Volume 11, pp. 115–193.
51. Marks, T.J.; Ratner, M.A. Design, synthesis, and properties of molecule-based assemblies with large second-order optical nonlinearities. *Angew. Chem. Int. Ed. Engl.* **1995**, *34*, 155–173. [CrossRef]

symmetry MDPI

Article

The Missing Relationship between the Miscibility of Chiral Dopants and the Microscopic Dynamics of Solvent Liquid Crystals: A Molecular Dynamics Study

Go Watanabe [1,2,*,†], Akane Yamazaki [1] and Jun Yoshida [3,*]

1 School of Science, Kitasato University, Sagamihara 252-0373, Japan
2 Kanagawa Institute of Industrial Science and Technology, Kawasaki 243-0435, Japan
3 Graduate School of Integrated Basic Sciences, Nihon University, Tokyo 156-8550, Japan
* Correspondence: go0325@kitasato-u.ac.jp (G.W.); yoshida.jun@nihon-u.ac.jp (J.Y.)
† Current address: School of Frontier Engineering, Kitasato University, Sagamihara 252-0373, Japan.

Abstract: Nematic liquid crystals (LCs) are known to undergo a phase transition to chiral nematic LCs possessing helices upon doping with enantiomeric molecules known as chiral dopants. The relationship between the helical pitch (p), the molar fraction (x), and the power of the chiral dopant to induce a helix in a nematic solvent (β_M) is expressed as $p = 1/(x{\cdot}\beta_M)$. The helical pitch is easily controlled by the concentration of the chiral dopant when the dopant molecule is miscible with the host nematic LC. However, it has not yet been clarified what the miscibility of the chiral dopant molecules with the nematic LCs depends. Therefore, we performed all-atom molecular dynamics (MD) simulations for the system composed of both Δ and Λ isomers of a chiral dopant molecule dispersed in a nematic LC and investigated the relationship between the microdynamics of the chiral molecules and their miscibility with the nematic solvent. The miscibility of the chiral dopant molecules with the LC solvent was found to correlate with the diffusion coefficient of the LC solvent. In the system where the chiral dopant molecules with high miscibility were added, the diffusion coefficient of the LC solvents was comparable to that of the system in which the chiral molecule was not doped. Furthermore, it was confirmed that more elongated chiral dopants were more miscible with the nematic solvent consisting of calamitic molecules, and that these dopant molecules did not have a significant effect on the diffusion behavior of the LC molecules.

Keywords: chiral nematic liquid crystals; molecular dynamics simulation; miscibility; diffusion coefficient

Citation: Watanabe, G.; Yamazaki, A.; Yoshida, J. The Missing Relationship between the Miscibility of Chiral Dopants and the Microscopic Dynamics of Solvent Liquid Crystals: A Molecular Dynamics Study. *Symmetry* **2023**, *15*, 1092. https://doi.org/10.3390/sym15051092

Academic Editor: Cosimo Cardellicchio

Received: 17 March 2023
Revised: 6 May 2023
Accepted: 11 May 2023
Published: 16 May 2023

1. Introduction

It is generally well known that a nematic liquid crystal (LC) undergoes a phase transition to a chiral nematic LC possessing a helix when doped with enantiomeric molecules known as chiral dopants [1–4]. The chiral nematic LC exhibits selective reflection of circularly polarized light corresponding to the helical pitch. The pitch of the induced helicoidal superstructure is usually within the visible wavelength range. As the helical pitch can be tuned by the type and concentration of the chiral dopants, chiral nematic LCs have the potential to be utilized for photonic applications in areas of tunable mirrorless lasing [5–7], reflective colored displays [8–10], energy-saving smart windows [11–13], and sensors [14,15].

The relationship between the helical pitch (p), the molar fraction (x), and the power of the chiral dopant used to induce a helix in a nematic solvent (helical twisting power, denoted as β_M) is expressed as follows: $p = 1/(x{\cdot}\beta_M)$. The sign of the β_M value indicates the helical direction, where positive and negative values correspond to P- and M-helical structures, respectively.

The rational prediction of β_M of the chiral dopant is associated with the precise control of the helical structure of chiral nematic LCs. Although the helix formation in chiral nematic LCs is well understood as a macroscopic physical phenomenon, predicting the values of β_M rationally remains a challenging task based on the dopant and LC molecules. Slight differences of the molecular structure of the dopant may sometimes lead to large differences in the value of β_M.

Unlike predicting the values of β_M of the chiral dopant, the helical pitch is easily controlled by the concentration of the chiral dopant when the dopant molecule is miscible with the host nematics. However, the key factor determining the miscibility of chiral dopant molecules with nematic LCs has not yet been clarified.

The molecular shape of the chiral dopant may be one possible factor affecting its miscibility with nematics. To clarify the relationship between the molecular shape and miscibility of the chiral dopant molecule, octahedral ruthenium complexes, especially [Ru(acac)$_3$] (acac = acetylacetonate ion), are suitable chiral dopants; they have rigid and C_3 symmetric central cores, and their shapes can be systematically changed by the introduction of elongated moieties to one of the acac ligands. In previous studies, the helical twisting powers of ruthenium complex dopants were investigated using elongated ligands (**Ru-trop**) [16] and simple bromo groups (**Ru-Br0** and **Ru-Br1**) for comparison [17] (Figure 1). **Ru-Br0** and **Ru-Br1** are miscible with *N*-(4-methoxybenzylidene)-4-butylaniline (MBBA), which is well-known to show a nematic LC phase at room temperature, at no more than 0.2 mol%, while no precipitation of the dopant was observed using polarized optical microscopy for ~4.0 mol% of **Ru-trop** in MBBA [16]. There was no significant difference in the viscosity among these chiral LC mixtures with low concentrations of the chiral dopants. The difference in the microscopic behavior of these ruthenium complexes may be the cause of the difference in miscibility, but it is challenging to confirm this experimentally.

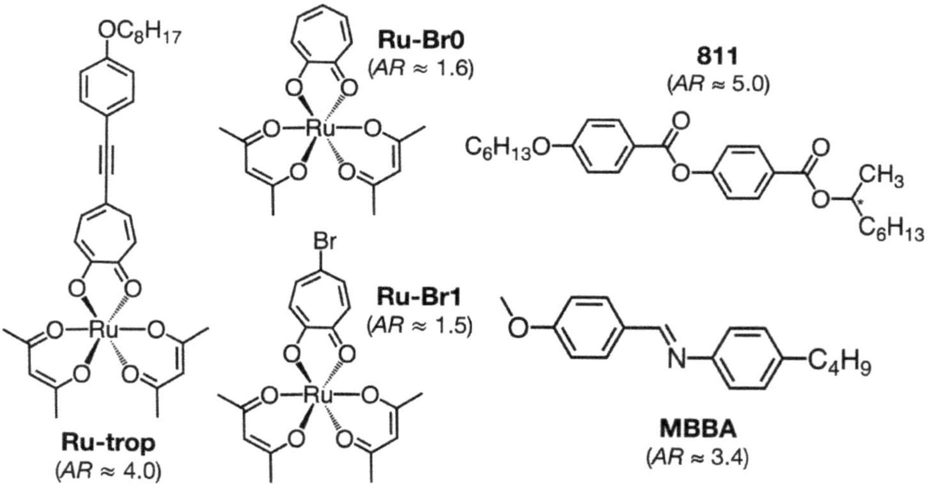

Figure 1. Molecular structures of Δ-isomers of ruthenium complexes (**Ru-trop**, **Ru-Br0**, and **Ru-Br1**) and an organic compound (**811**) used as a chiral dopant and the nematic LC (MBBA) used as a host. The values in the brackets for the ruthenium complexes and organic compounds represent the aspect ratio of the molecular length parallel to the C_2 axis to that perpendicular to the C_2 axis and that of the molecular long axis to the short axis, respectively.

We herein present new insights into the dynamics of chiral dopants at the atomic level, which are related to their macroscopic miscibility with nematic solvents. To achieve this, we conducted all-atom molecular dynamics (MD) simulations.

2. Models and Methods

In this study, all-atom MD simulations were performed using the program GROMACS 2016.3. The simulated racemic system consisted of both Δ- and Λ-enantiomers of a chiral dopant molecule dispersed in the nematic LC molecules, as previously proposed. This approach was taken because it is currently not realistic to perform all-atom MD simulations for chiral nematic LC systems exhibiting a helical structure with a sub-micrometer pitch [18]. While this approach cannot reproduce the macroscopic helical structure, it is sufficient for investigating the microscopic behavior of the LC molecules and the dopants. Three types of chiral ruthenium complex compounds, **Ru-Br0**, **Ru-Br1**, and **Ru-trop**, with different solubilities in the nematic solvents, and one chiral organic compound, **811**, with high miscibility in the nematics, as shown in Figure 1 were selected as the chiral dopants. The **811** compound has a mesogen core, which is expected to be miscible with a mesogenic compound. The selected LC molecule was MBBA, which is well known as a room temperature LC. The diffusion coefficients of pure MBBA obtained from the experiments and the MD simulation were found to be close each other, as described later. The **811** compound is miscible with MBBA, with a miscibility of at least 20 mol%. For each simulated system, the concentration of the dopant was set to ~0.1 mol%, and the number of molecules was listed in Table 1. The number of LC molecules were set to approximately 2000, as the system with this number of molecules can accurately reproduce the nematic-isotropic phase transition.

Table 1. Number of molecules in the five systems for pure MBBA, **N-Ru-Br0**, **N-Ru-Br1**, **N-Ru-trop**, and **N-811**.

System	LC Molecules	Chiral Dopant Molecule
Pure MBBA	2016	0
N-Ru-Br0	2004	2
N-Ru-Br1	2006	2
N-Ru-trop	2002	2
N-811	2003	2

The initial structure was constructed according to the methodology previously proposed [18]. In the initial structure of the pure MBBA, 2016 molecules were randomly placed without overlapping in a cubic periodic boundary box with an 11.4 nm side length. For the system in which Δ- and Λ-isomers were added to MBBA molecules, these isomers were placed not to interact with each other at the initial state.

The generalized Amber force field [19] parameters were used for the force field parameters for MBBA and chiral dopant molecules, while the universal force field (UFF) [20] parameters were applied to calculate the bond-stretching and bond-angle bending interactions, including interatomic bonds between ruthenium and its neighboring oxygen atoms. The partial atomic charges of the molecules were calculated using the restrained electrostatic potential (RESP) methodology [21] based on density functional theory (DFT) calculations (B3LYP/LANL2DZ for ruthenium atom and B3LYP/6-31G* for all other atoms) with the GAUSSIAN 09 revision E01 program.

The simulation procedure proposed in the previous study [18] involved pre-equilibration runs after the steepest energy minimization, which were performed as follows: 5 ns at 250 K and 1 bar, 1 ns at 350 K and 1 bar, and 5 ns at 300 K and 1 bar. During the pre-equilibration runs, the temperature and pressure were kept constant using a Berendsen thermostat and barostat [22] with relaxation times of 0.2 and 2.0 ps, respectively. The equilibration run was carried out for at least 300 ns at 300 K and 1 bar using a Nosé-Hoover thermostat [23–25] and Parrinello-Rahman barostat [26] with relaxation times of 1.0 and 5.0 ps, respectively. The time step was set to 2 fs, since all bonds connected to hydrogen atoms were constrained with the LINCS algorithm [27]. The long-range Coulomb interactions were calculated using the smooth particle-mesh Ewald method [28] with a grid spacing of 0.30 nm. The real space cutoff for both Coulomb and van der Waals interactions was 1.4 nm.

3. Results

Figure 2 shows the MD simulation snapshots of the simulated system, pure MBBA, **N-Ru-Br0**, **N-Ru-Br1**, **N-Ru-trop**, and **N-811**. In all the systems, Δ- and Λ-isomers were positioned away from each other, and no aggregations were formed. The MBBA molecules appeared to be aligned in one arbitrary direction in the snapshots, but the order parameter for the molecules must be calculated to understand the degree of molecular ordering.

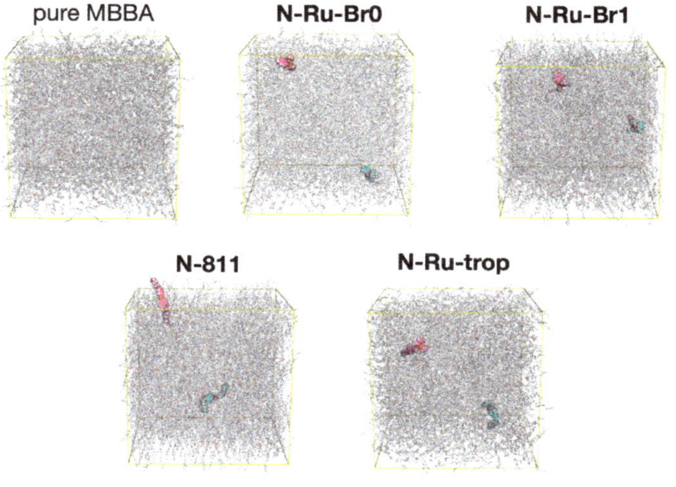

Figure 2. MD simulation snapshots of pure MBBA, **N-Ru-Br0**, **N-Ru-Br1**, **N-Ru-trop**, and **N-811** after the equilibration MD run. The carbon atoms of Δ- and Λ-isomers are colored in magenta and cyan, respectively. MBBA molecules are illustrated by sticks.

The order parameter S was estimated as the largest positive eigenvalue of the order parameter tensor $\boldsymbol{P}$, as expressed below:

$$P_{\alpha\beta} = \frac{1}{N}\sum_{i=1}^{N}\frac{1}{2}\left(3n_{i\alpha}n_{i\beta} - \delta_{\alpha\beta}\right) \tag{1}$$

where N is the total number of the LC molecules, subscripts α and β represent the coordinates x, y, and z, and $\boldsymbol{n}_i$ is the normalized long axis vector of the ith molecule. The long axis was defined as the vector connecting two end carbon atoms of the rigid aromatic core, as referred to in the previous study [29]. The time dependence of the nematic order parameter for the MBBA molecules was analyzed for each simulated system, as shown in Figure 3.

For the pure MBBA system, the order parameter averaged during the last 50 ns of the equilibration MD run was 0.57 ± 0.01 at 300 K. The system obtained from the MD simulations could exhibit the characteristics of bulk nematics, as the simulated order parameter is consistent with the experimental value for MBBA [30].

To confirm the validity of the simulation model and conditions, the temperature dependence of the order parameter for the pure MBBA system was also investigated, as shown in Figure 4. The order parameter at each temperature was the average of the last 50 ns of the equilibration MD run. Figure 4 indicates that the phase transition temperature for the MD simulation was close to the experimental value of 318 K.

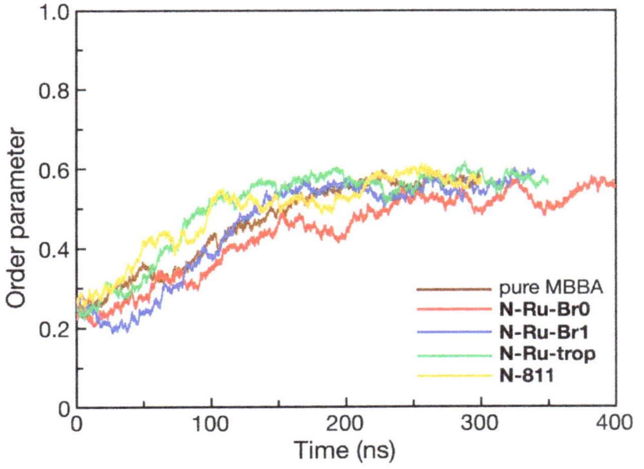

Figure 3. Time dependence of the nematic order parameter for pure MBBA, **N-Ru-Br0**, **N-Ru-Br1**, **N-Ru-trop**, and **N-811**. All the systems reached equilibration in the final 100 ns.

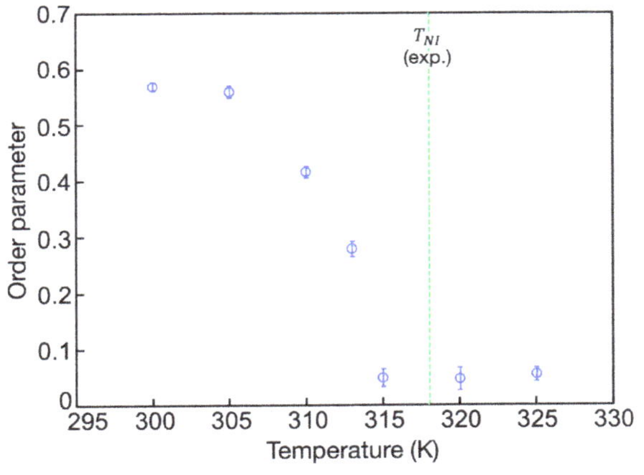

Figure 4. Temperature dependence of the nematic order parameter for pure MBBA. The values of the order parameter above 315 K were nearly zero, indicating that these systems were isotropic.

The order parameters averaged during the last 50 ns of the equilibration MD run were 0.57 ± 0.01, 0.54 ± 0.02, 0.57 ± 0.02, 0.58 ± 0.02, and 0.58 ± 0.02 for pure MBBA, **N-Ru-Br0**, **N-Ru-Br1**, **N-Ru-trop**, and **N-811**, respectively. These values indicate that these systems possessed the same degree of orientational order as that of the pure MBBA system.

The radial distribution functions are typically analyzed to investigate whether the simulated system has positional order. The radial distribution function (RDF) of the LC molecules, $g(r)$, is defined as:

$$g(r) = \frac{1}{4\pi r^2 \Delta r} \frac{V}{N} \frac{1}{N} \sum_{i=1}^{N} \Delta N_i(r) \quad (2)$$

where V is the total volume of the system, N is the total number of the LC molecules, $\Delta N_i(r)$ is the number of the LC molecules in the shell at a distance r from the ith molecule, and Δr is the thickness of the shell. As shown in Figure S1, the RDFs for all systems exhibit similar curves, do not exhibit discrete sharp peaks, and asymptotically converge to 1. These characteristics of the RDFs are typical of those in the liquid state. It is shown that the LC molecules in the system were arranged in a disorderly manner without long-positional order. However, each of the RDFs had the first broad peak at approximately 0.5 nm, which is comparable to the length of the short axis of the MBBA molecule. This suggests a parallel arrangement with short-range order.

The diffusion properties of LC can be analyzed through all-atom MD simulation, providing important insights from a microscopic point of view [31]. The mean square displacements (MSDs) of the host MBBA molecules were analyzed, and the diffusion coefficients were calculated using Einstein's relation defined by following equation:

$$D = \frac{\langle |\boldsymbol{r}_i(t_0 + t) - \boldsymbol{r}_i(t_0)|^2 \rangle_i}{6t} = \frac{a}{6} \tag{3}$$

where $\boldsymbol{r}_i(t_0)$ is the molecular center-of-mass position at time t_0, and a is the slope of the linear fit of the MSD. Figure 5 indicates that the MBBA molecules in all the systems had significant fluidity. The diffusion coefficients of the MBBA molecules listed in Table 2 were of the same order of the magnitude as those reported in previous experiments [32]. Although the diffusion coefficients of the pure MBBA system, **N-Ru-trop**, and **N-811** were close to each other, the values were higher than those of **N-Ru-Br0** and **N-Ru-Br1**. Therefore, the more soluble the chiral dopant was in the nematics, the larger the diffusivity of the host LC molecules. This indicates that there is a correlation between the miscibility of the chiral dopant in the host LC solvent and the diffusion coefficient of the LC molecules.

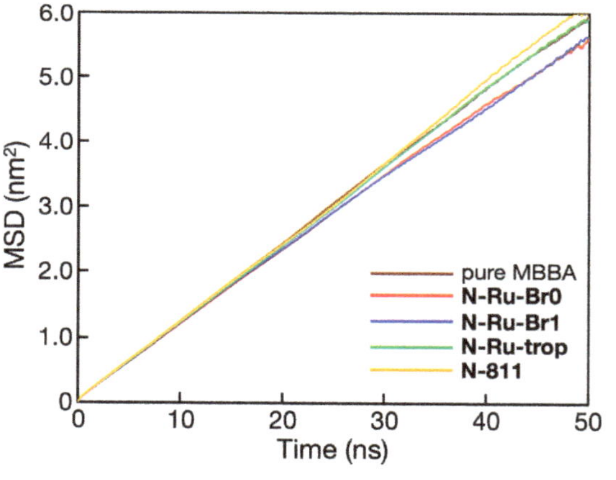

Figure 5. MSDs of MBBA molecules during the last 50 ns of equilibration runs for pure MBBA, **N-Ru-Br0**, **N-Ru-Br1**, **N-Ru-trop**, and **N-811**. The slope of the line was calculated using least square fitting between 15 ns and 45 ns.

Table 2. Diffusion coefficients (m^2/s) of MBBA in the five systems for pure MBBA, **N-Ru-Br0**, **N-Ru-Br1**, **N-Ru-trop**, and **N-811**.

	Pure MBBA	N-Ru-Br0	N-Ru-Br1	N-Ru-trop	N-811
D (m^2/s)	1.99 ± 0.01	1.87 ± 0.02	1.84 ± 0.02	2.00 ± 0.02	2.06 ± 0.04

Total interaction energy, which is defined as the sum of the van der Waal and Coulomb interaction energies, was calculated between the Δ-isomer of the chiral dopant molecule and the surrounding LC molecules for each system, and the time dependence is shown in Figure 6. There is little difference among the systems irrespective of the chiral ruthenium complexes doped. The total interaction energies for pure MBBA and **N-811** were comparable and lower than those for the other three systems. No significant difference in the Coulomb interaction energy was seen among all the systems as shown in Figure S2. The van der Waals interaction energies of **Ru-Br0** and **Ru-Br1**, which are less miscible with LC, were larger than those of **Ru-trop** and **811** with high miscibility. This suggests that the van der Waals interaction energy between the chiral dopant and the host LC molecules affect its miscibility with the LC solvents.

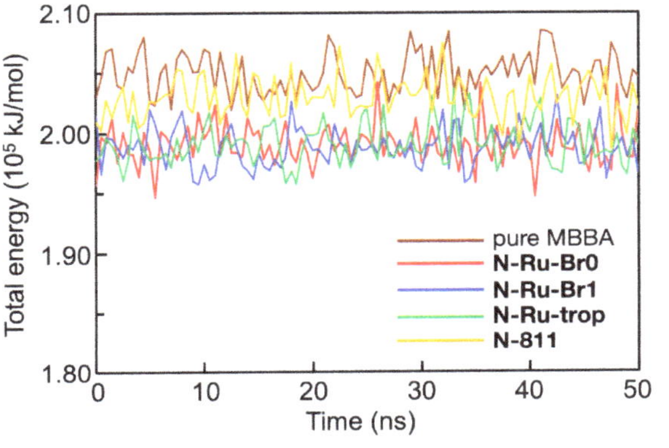

Figure 6. Time dependences of total interaction energies between the Δ-isomer of the chiral dopant molecule and the MBBA molecules for pure MBBA, **N-Ru-Br0**, **N-Ru-Br1**, **N-Ru-trop**, and **N-811** during the last 50 ns of the equilibration runs.

The MD simulations were performed for the systems with twice the concentration of **Ru-Br0** and **Ru-trop**. The time dependence of the order parameter of MBBA molecules for **N-Ru-Br0_2** with 0.2 mol% of **Ru-Br0** and **N-Ru-trop_2** with 0.2 mol% of **Ru-trop** is shown in Figure S4. The average order parameters of the LC molecules in the last 50 ns of the equilibration MD runs were 0.56 ± 0.01 and 0.59 ± 0.01, respectively. These values were similar to those of **N-Ru-Br0** and **N-Ru-trop**. From the slope of the MSD shown in Figure S7, the diffusion coefficients of MBBA molecules were estimated as $(1.74 \pm 0.01) \times 10^{-11}$ m^2/s and $(1.88 \pm 0.01) \times 10^{-11}$ m^2/s for **N-Ru-Br0_2** and **N-Ru-trop_2**, respectively. In both **N-Ru-Br0_2** and **N-Ru-trop_2**, compared to **N-Ru-Br0** and **N-Ru-trop** in which the concentrations of the chiral dopant were 0.1 mol%, the diffusion coefficients of the LC molecules were slightly smaller. Since the diffusion coefficient of the LC molecules of **N-Ru-trop_2** was significantly larger than that of **N-Ru-Br0_2**, the correlation between the diffusion property of the LC solvent and the miscibility of the chiral dopant was independent of its concentration. Figure S3 shows the time dependence of the total interaction energies, the van der Waals interaction energies, and Coulomb interaction energies between the Δ-isomer of the chiral dopant molecule and the LC molecules. There were no significant differences in the interaction energies depending on the concentration of chiral dopants.

In the initial structure of the systems described above, the chiral dopant molecules were dispersed into the LC molecules and did not form any aggregates during the equilibration runs. It is crucial to determine whether the dopants are dispersed in nematic liquid crystals. To confirm the dispersibility of the dopants in detail, MD simulations were performed for the systems in which a pair of Δ- and Λ-isomers was placed in the LC solvent as an aggregated initial structure. The simulated systems are denoted as **N-Ru-Br0_agg** and **N-Ru-trop_agg**, containing MBBA molecules and **Ru-Br0** and **Ru-trop**, respectively. During the pre-equilibrium and the following equilibration runs, a pair of Δ- and Λ-isomers was restrained in the initial position by applying the harmonic potential with a force constant of 1000 $kJ/mol/nm^2$. After 200 ns, during which the system reached the equilibrated state

(Figure S4), an equilibration run without any restraints was performed for 400 ns (Figure S6). In both **N-Ru-Br0_agg** and **N-Ru-trop_agg**, the aggregation was not stable, and the two isomers individually diffused in the nematic solvent (Figure 7). The distance between the isomers was approximately 4.0 nm, as shown in Figure S9. The MSDs of MBBA molecules for the last 50 ns of the equilibration run with no restraints were analyzed, as shown in Figure S8, and the diffusion coefficients were estimated as $(1.84 \pm 0.02) \times 10^{-11}$ m^2/s and $(1.95 \pm 0.01) \times 10^{-11}$ m^2/s for **N-Ru-Br0_agg** and **N-Ru-trop_agg**, respectively. The LC molecules in **N-Ru-trop_agg** diffused slightly faster than those in **N-Ru-Br0_agg**. The value of the diffusion coefficient of the LC molecules in **N-Ru-trop_agg** was nearly the same as those in pure MBBA and **N-Ru-trop.** This indicates that the systems in which the aggregate of Δ- and Λ-isomers was initially placed in the LC solvents exhibited the same behavior as those in which the isomers were initially separated. Since the difference in the formation of the chiral molecules in the initial structure had little effect on the physical properties, the validity of the initial structures of **N-Ru-Br0**, **N-Ru-Br1**, **N-Ru-trop**, and **N-811** was also confirmed.

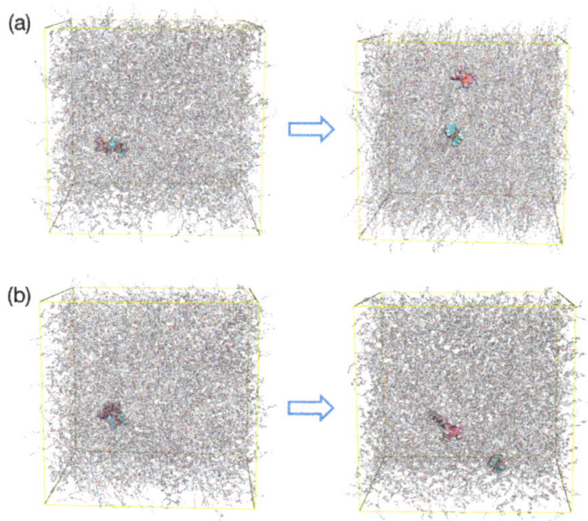

Figure 7. MD simulation snapshots of (**a**) **N-Ru-Br0_agg** and (**b**) **N-Ru-trop_agg** at 0 ns (**left**) and at 400 ns (**right**) of the equilibration run without any restraints. The carbon atoms of Δ- and Λ-isomers are colored in magenta and cyan, respectively. MBBA molecules are illustrated by sticks.

4. Summary

In this study, all-atom MD simulations for the racemic systems in which both the Δ- and Λ-enantiomers of the chiral dopant molecule were doped into the nematic LC molecul, for **Ru-Br0**, **Ru-Br1**, **Ru-trop**, and **811**. By investigating the microscopic behavior of the chiral dopant and LC molecules, we aimed to understand the factor that causes the difference in the miscibility of the chiral dopant analogues in the nematic solvents.

The nematic order parameters of LC molecules for all the systems into which the chiral dopant molecules were doped were the same as that of the pure LC system. At low concentrations of the chiral dopants, such as 0.1 mol%, they had little effect on the orientational order of the surrounding LC molecules.

By analyzing the MSDs and diffusion coefficients of the LC molecules, the miscibility of the chiral dopant in the host LC solvent and the diffusion behavior of the LC molecules were found to be correlated. In the system to which the chiral dopant molecules with high miscibility were added, the diffusion coefficient of the LC solvents was comparable to that of the system without the chiral molecule. This indicates that the diffusion property, i.e., the viscosity of the LC molecules, was less affected by the chiral dopant molecule with higher miscibility.

The analysis of the interaction energies between the chiral dopant molecule and LC molecules showed that the van der Waals interaction energies were correlated with the miscibility of the dopant molecule. The molecular shapes of **Ru-trop** and **811**, which are highly miscible with MBBA solvent, are more elongated compared to **Ru-Br0** and **Ru-Br1**. The aspect ratio of the molecular length parallel to the C_2 axis to that perpendicular to the C_2 axis for each ruthenium complex molecule was approximately 1.6, 1.5, and 4.0 for **Ru-Br0**, **Ru-Br1**, and **Ru-trop**, respectively. These molecules are derivatives with the same rigid octahedral ruthenium core, but the shape of **Ru-trop** is more rod-like and anisotropic compared to that of **Ru-Br0** and **Ru-Br1**. This indicates that more elongated chiral dopants were more miscible with the nematic solvent consisting of calamitic molecules, and that those dopant molecules had little influence on the diffusion behavior of the nematic LC molecules.

Moreover, it is of interest to investigate how the sign of dielectric anisotropy and the abundance of the ion in the LC molecules influence on their diffusion behaviors. To address these issues, we plan to conduct additional MD simulations using different solvents in the future.

Several theoretical and computational methods to have been developed to accurately calculate the thermodynamic and structural properties of solvation [33]. However, predicting the miscibility of organic molecules in anisotropic fluids remains challenging, even though solvation free energies for small gas molecules have been calculated by determining the free energy change for transferring each gas molecule from an ideal gas vapor phase to bulk LC [34].

Performing MD simulations using the model and procedure presented in this paper may enable us to predict the miscibility of a newly designed dopant molecule in nematic LC solvents before synthesizing the dopant. This approach could help facilitate the development of novel chemical compounds more efficiently than ever before.

Supplementary Materials: The following supporting information can be downloaded at https://www.mdpi.com/article/10.3390/sym15051092/s1.

Author Contributions: A.Y. and G.W. performed MD simulations and analysis. J.Y. and G.W. co-wrote manuscript. All authors have read and agreed to the published version of the manuscript.

Funding: The computations were partially performed at the Research Center for Computational Science, Okazaki, Japan (Project: 21-IMS-C043, 22-IMS-C043). This work was supported by the Grant-in-Aid for Scientific Research (B) (JP19H02537), Grant-in-Aid for Scientific Research (C) (JP19K05508), Grant-in-Aid for Scientific Research on Innovative Areas (JP19H05718), and Grant-in-Aid for Challenging Research (Exploratory) (JP22K18953).

Data Availability Statement: The data presented in this study are available on request from the corresponding author (G.W.).

Conflicts of Interest: The authors declare no conflict of interest.

References

1. Solladié, G.; Zimmermann, R.G. Liquid Crystals: A Tool for Studies on Chirality. *Angew. Chem. Int. Ed.* **1984**, *23*, 348–362. [CrossRef]
2. de Gennes, P.G.; Prost, J. *The Physics of Liquid Crystals*, 2nd ed.; Oxford University Press: Oxford, UK, 1993.
3. Kitzerow, H.-S.; Bahr, C. (Eds.) *Chirality in Liquid Crystals*; Springer: New York, NY, USA, 2001.
4. Goodby, J.W.; Collings, P.J.; Kato, T.; Tschierske, C.; Gleeson, H.; Raynes, P. (Eds.) *Handbook of Liquid Crystals*; Wiley-VCH Verlag GmbH & Co. KGaA: Weinheim, Germany, 2014.
5. Finkelmann, H.; Kim, S.T.; Muñoz, A.; Palffy-Muhoray, P.; Taheri, B. Tunable mirrorless lasing in cholesteric liquid crystalline elastomers. *Adv. Mater.* **2001**, *13*, 1069–1072. [CrossRef]
6. Matsui, T.; Ozaki, R.; Funamoto, K.; Ozaki, M.; Yoshino, K. Flexible mirrorless laser based on a free-standing film of photopolymerized cholesteric liquid crystal. *Appl. Phys. Lett.* **2002**, *81*, 3741–3743. [CrossRef]

7. Araoka, F.; Shin, K.C.; Takanishi, Y.; Ishikawa, K.; Takezoe, H.; Zhu, Z.; Swager, T.M. How doping a cholesteric liquid crystal with polymeric dye improves an order parameter and makes possible low threshold lasing. *J. Appl. Phys.* **2003**, *94*, 279–283. [CrossRef]
8. Yang, D.K.; Doane, J.W.; Yaniv, Z.; Glasser, J. Cholesteric Reflective Display: Drive Scheme and Contrast. *Appl. Phys. Lett.* **1994**, *64*, 1905–1907. [CrossRef]
9. Tamaoki, N. Cholesteric Liquid Crystals for Color Information Technology. *Adv. Mater.* **2001**, *13*, 1135–1147. [CrossRef]
10. Hsiao, Y.-C.; Yeh, E.-R.; Lee, W. Advanced Color-Reflective Dual-Frequency Cholesteric Liquid Crystal Displays and the Driving Matrix. *Mol. Cryst. Liq. Cryst.* **2017**, *644*, 12–18. [CrossRef]
11. Yu, B.H.; Huh, J.W.; Heo, J.; Yoon, T.H. Simultaneous Control of Haze and Transmittance Using a Dye-Doped Cholesteric Liquid Crystal Cell. *Liq. Cryst.* **2015**, *42*, 1460–1464. [CrossRef]
12. Oh, S.-W.; Kim, S.-H.; Baek, J.-M.; Yoon, T.-H. Optical and Thermal Switching of Liquid Crystals for Self-Shading Windows. *Adv. Sustain. Syst.* **2018**, *2*, 1700164. [CrossRef]
13. Li, C.; Tseng, H.; Chen, C.; Wang, C.; Jau, H.; Wu, Y.; Hsu, W.; Lin, T.H. Versatile Energy-Saving Smart Glass Based on Tristable Cholesteric Liquid Crystals. *ACS Appl. Energy Mater.* **2020**, *3*, 7601–7609. [CrossRef]
14. Han, Y.; Pacheco, K.; Bastiaansen, C.W.M.; Broer, D.J.; Sijbesma, R.P. Optical Monitoring of Gases with Cholesteric Liquid Crystals. *J. Am. Chem. Soc.* **2010**, *132*, 2961–2967. [CrossRef] [PubMed]
15. Zhang, W.; Schenning, A.P.H.J.; Kragt, A.J.J.; Zhou, G.; de Haan, L.T. Thermochromic Multicolored Photonic Coatings with Light Polarization- and Structural Color-Dependent Changes. *ACS Appl. Polym. Mater.* **2022**, *4*, 537–545. [CrossRef]
16. Yoshida, J.; Tamura, S.; Watanabe, G.; Kasahara, Y.; Yuge, H. "Colored" Inorganic Dopants for Inducing Liquid Crystal Chiral Nematic and Blue Phases: Monitoring of Dopant-Host Interaction by Raman Spectroscopy. *Chem. Commun.* **2017**, *53*, 5103–5106. [CrossRef]
17. Yoshida, J.; Tamura, S.; Hoshino, K.; Yuge, H.; Sato, H.; Yamazaki, A.; Yoneda, S.; Watanabe, G. Comprehensive Understanding of Host- and Guest-Dependent Helix Inversion in Chiral Nematic Liquid Crystals: Experimental and Molecular Dynamics Simulation Study. *J. Phys. Chem. B* **2018**, *122*, 10615–10626. [CrossRef]
18. Yoshida, J.; Watanabe, G. Molecular Dynamics Approach for Predicting Helical Twisting Powers of Metal Complex Dopants in Nematic Solvents. *J. Phys. Chem. B* **2016**, *120*, 6858–6864.
19. Wang, J.; Wolf, R.M.; Caldwell, J.W.; Kollman, P.A.; Case, D.A. Development and Testing of a General Amber Force Field. *J. Comput. Chem.* **2004**, *25*, 1157–1174. [CrossRef] [PubMed]
20. Rappe, A.K.; Casewit, C.J.; Colwell, K.S.; Goddard III, W.A.; Skiff, W.M. UFF, a Full Periodic Table Force Field for Molecular Mechanics and Molecular Dynamics Simulations. *J. Am. Chem. Soc.* **1992**, *114*, 10024–10035. [CrossRef]
21. Bayly, C.I.; Cieplak, P.; Cornell, W.; Kollman, P.A. A Well-Behaved Electrostatic Potential Based Method Using Charge Restraints for Deriving Atomic Charges: The RESP Model. *J. Phys. Chem.* **1993**, *97*, 10269–10280. [CrossRef]
22. Berendsen, H.J.C.; Postma, J.P.M.; van Gunsteren, W.F.; DiNola, A.; Haak, J.R. Molecular Dynamics with Coupling to an External Bath. *J. Chem. Phys.* **1984**, *81*, 3684–3690. [CrossRef]
23. Nosé, S. A Molecular Dynamics Method for Simulations in the Canonical Ensemble. *Mol. Phys.* **1984**, *52*, 255–268. [CrossRef]
24. Nosé, S. A unified formulation of the constant temperature molecular dynamics methods. *J. Chem. Phys.* **1984**, *81*, 511–519. [CrossRef]
25. Hoover, W.G. Canonical dynamics: Equilibrium phase-space distributions. *Phys. Rev. A* **1985**, *31*, 1695–1697. [CrossRef] [PubMed]
26. Parrinello, M.; Rahman, A. Polymorphic Transitions in Single Crystals: A New Molecular Dynamics Method. *J. Appl. Phys.* **1981**, *52*, 7182–7190. [CrossRef]
27. Hess, B.; Bekker, H.; Berendsen, H.J.C.; Fraaije, J.G.E.M. LINCS: A linear constraint solver for molecular simulations. *J. Comput. Chem.* **1997**, *18*, 1463–1472. [CrossRef]
28. Essmann, U.; Perera, L.; Berkowitz, M.L.; Darden, T.; Lee, H.; Pedersen, L.G. A smooth particle mesh Ewald method. *J. Chem. Phys.* **1995**, *103*, 8577–8593. [CrossRef]
29. Watanabe, G.; Saito, J.; Kato, N.; Tabe, Y. Orientational correlations in two-dimensional liquid crystals studied by molecular dynamics simulation. *J. Chem. Phys.* **2011**, *134*, 054513. [CrossRef]
30. Subramhanyam, H.S.; Prabha, C.S.; Krishnamurti, D. Optical Anisotropy of Nematic Compounds. *Mol. Cryst. Liq. Cryst.* **1974**, *28*, 201–215. [CrossRef]
31. Watanabe, G.; Saito, J.; Fujita, Y.; Tabe, Y. Molecular Dynamics Simulation Study of Two-Dimensional Diffusion Behavior in Smectic Liquid Crystalline Monolayers. *J. Phys. Soc. Jpn.* **2013**, *82*, 084603. [CrossRef]
32. Ohta, K.; Terazima, M.; Hirota, N. Diffusion Process of a Liquid Crystal Probed by the Transient Grating Method. *Bull. Chem. Soc. Jpn.* **1995**, *68*, 2809–2815. [CrossRef]
33. Ratkova, E.L.; Palmer, D.S.; Fedorov, M.V. Solvation Thermodynamics of Organic Molecules by the Molecular Integral Equation Theory: Approaching Chemical Accuracy. *Chem. Rev.* **2015**, *115*, 6312–6356. [CrossRef]
34. Sheavly, J.K.; Gold, J.I.; Mavrikakis, M.; Van Lehn, R.C. Molecular simulations of analyte partitioning and diffusion in liquid crystal sensors. *Mol. Syst. Des. Eng.* **2020**, *5*, 304–316. [CrossRef]

symmetry

MDPI

Article

New Diluter Solubilized in Liquid Crystal Compounds for High Stability and Fast Response Speed Liquid Crystal Displays

Masanobu Mizusaki [1,*], Kazuo Okamoto [2] and Toshihiro Shibata [3]

1 Sharp Display Technology Corporation, Development Group, Tenri 632-8567, Japan
2 Organo Science Co., Ltd., Omaezaki 437-1617, Japan
3 Chiracol Co., Ltd., Saitama 331-0823, Japan
* Correspondence: mizusaki.masanobu@sharp.co.jp; Tel.: +81-50-5433-4691

Abstract: Diluters play a crucial role for reducing the rotational viscosity of liquid crystal (LC) materials for improving the response property of LC displays (LCDs). However, conventional diluters tend to deteriorate relatively easily under ultraviolet (UV) exposure, due to existence of vinyl groups. Hence, it was difficult to prepare polymer layers in the LC cells by exposing UV for polymerization regarding polymer-sustained vertically-aligned (PS-VA) and polyimide-free (PI-free) technologies. In this study, however, we succeeded to develop a high stable diluter, which shows a symmetrical structure and carries fluorinated groups for stable nematic phase and high affinity with other fluorinated LC compounds. With the use of the symmetrical-structured diluter, we developed the PS-VA and the PI-free in-plane switching cells with a fast response rate and high voltage holding ratio, by combining a reactive monomer including an azobenzene moiety.

Keywords: liquid crystal display; liquid crystal mixture; diluter; stability; polymer-sustained vertically aligned (PS-VA); PI-free

Citation: Mizusaki, M.; Okamoto, K.; Shibata, T. New Diluter Solubilized in Liquid Crystal Compounds for High Stability and Fast Response Speed Liquid Crystal Displays. *Symmetry* **2022**, *14*, 1620. https://doi.org/10.3390/sym14081620

Academic Editor: Rui Tamura

Received: 8 July 2022
Accepted: 4 August 2022
Published: 6 August 2022

1. Introduction

Liquid crystal displays (LCDs), which show a high contrast ratio and utilize a fast response speed have become significantly important in our lives. In particular, polymer-sustained vertically-aligned (PS-VA) LC is useful for large sized LCDs [1–3]. The PS-VA LCD shows high contrast ratio, fast response speed, and low driving voltage because LC molecules exhibit a slightly inclined pretilt angle from homeotropic alignment [1,2]. In addition, polyimide (PI)-free VA and in-plane switching (IPS) technologies are also paid much attention because the PI-free LCDs are expected to prepare extremely narrow picture frame LCDs [4–8]. In these techniques, ultraviolet (UV) light is generally exposed to prepare polymer layers by polymerization of reactive monomers in the LC layer. However, the UV exposure process for preparation of the polymer layer from the reactive monomers often induces serious image sticking [2,9]. One of causes for the image sticking is presumed to derive from a photo-chemical reaction between previous diluters carrying a vinyl group and the reactive monomers [10,11]. In the case that the photo-chemical reaction takes place in the LC layer, the image sticking tends to become observable. Hence, it is significantly important to develop new diluters for preventing the unexpected photo-chemical reaction. For this purpose, we developed a novel diluter with including two cyclohexane groups and two saturated alkyl groups (R_1-CyFF-Cy-R_2 (R_1, R_2: saturated alkyl groups)), which is depicted in Figure 1 [12]. We found that the number of saturated alkyl groups of R_1 and R_2 should be same for stable nematic phase. This fact indicates a symmetrical structure of the diluter is suitable for the LC materials. In addition, we prepared the LC mixtures, which are constructed from the LC materials and the reactive monomer, for the PS-VA and PI-free LC cells. In this study, we produced the PS-VA and PI-free IPS LC cells using the LC mixtures including the symmetrical diluter and various reactive monomers, and evaluated voltage holding ratio (VHR). The VHR is generally known as the image sticking parameter [13–16]. We confirmed that the combination of a newly developed diluter, and the reactive monomer carrying an azobenzene moiety exhibited high alignment stability and high VHR.

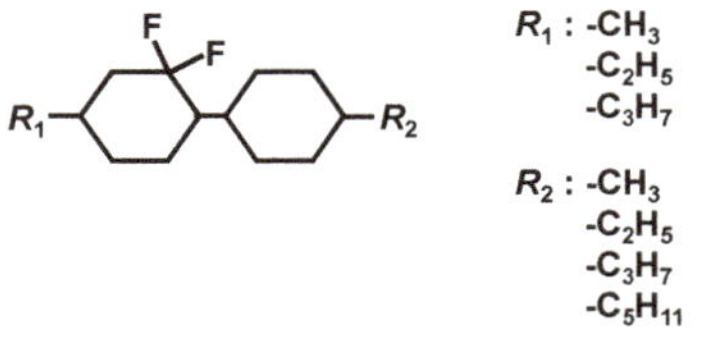

Figure 1. Chemical structure of the diluters R_1-CyFF-Cy-R_2 (R_1 and R_2 are saturated alkyl chains).

2. Experimental

2.1. Preparation of a New Diluter C3-CyFF-CyC3

As one of diluters, C3-CyFF-Cy-C3 is focused here, and we show a preparation procedure in the following Scheme 1. As the first step, a dehalogenation reaction between the compounds (1) and (2) prepared the compound (3). Then, the deoxofluorination reaction was performed by using diethylaminosulfur trifluoride (DAST) shown in the following scheme [12]. The GC-MS chart of the final compound is shown in Figure 2. The peak m/z 286 is observed, which is equal to the molecular weight of C3-CyFF-Cy-C3. Thus, we confirm that the diluter C3-CyFF-Cy-C3 is successfully obtained.

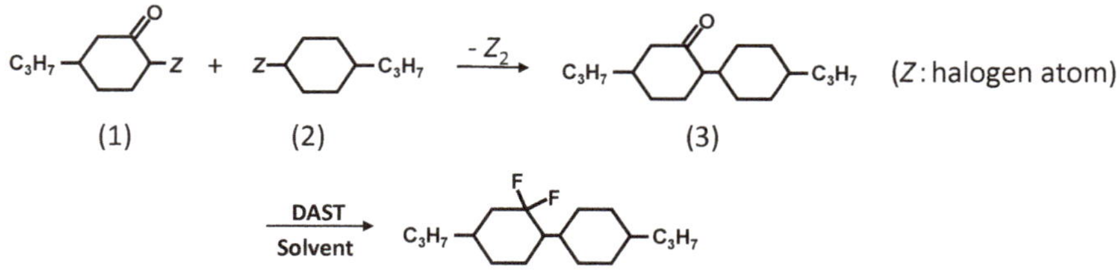

Scheme 1. Synthetic scheme of C3-CyFF-Cy-C3.

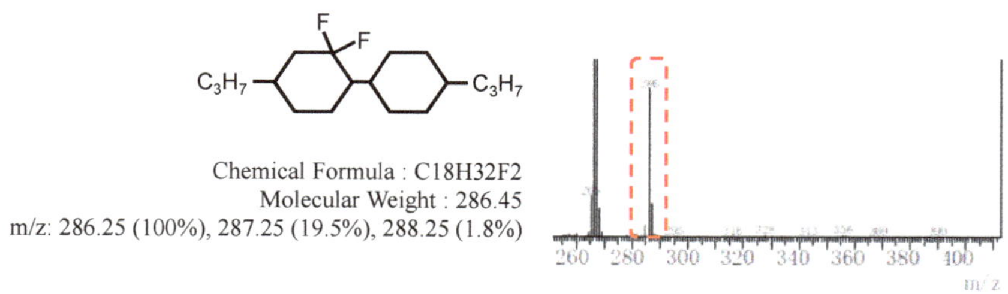

Figure 2. GC-MS chart of C3-CyFF-Cy-C3.

2.2. DSC Evaluation

DSC evaluation of the diluters R_1-CyFF-Cy-R_2 are performed by DSC7000X (Hitachi High Tech Science Corporation, Japan). The weight of samples was in the range of approximately 3.5–4.0 mg. The samples were encapsulated in Al pans, and were heated and cooled in dry nitrogen atmosphere. Measurements were carried out at 10.0 °C/min. The samples were heated from room temperature to 100 °C and cooled back to −50 °C at the same rate, and heated up to 100 °C again. The heat/cool cycles were repeated three times. The transition temperatures for various phase changes were obtained from endothermic peak of enthalpy, and determined from the third running.

2.3. Evaluation of the LC Cells

The LC cells (PS-VA and PI-free IPS cells) were prepared described in the following sections. The evaluation methods of each LC cell are described in the following sections.

The VHR measurements were carried out in a 70 °C controlled bath with a 6254-type VHR meter developed by Toyo Technica. The VHR was evaluated during an open-circuit period of 16.61 ms after ±5 V application.

A voltage-dependent transmittance curve (V-T curve) of the LC cells was measured at 25 °C with a Photal 5200 electro-optical evaluation system manufactured by Otsuka Electronics Corporation.

Response times, turn-on (τ_{on}) and turn-off (τ_{off}) times, are determined as transient times from 10% to 90% and from 90% to 10% of the maximum transmittance, respectively. These were determined by Photal 5200 electro-optical evaluation system.

Polarized optical microscopic (POM) images were observed by an ECLIPSE E600 POL system (Nikon Corp) in 25 °C controlled atmosphere.

3. New Developed Diluters

The newly developed diluters R_1-CyFF-Cy-R_2 illustrated in Figure 1 would be expected low rotational viscosity and low threshold voltage (Vth) because the diluters would show weak interaction with other LC compounds [12]. Most LC compounds have phenyl/phenylene rings, which usually interact with other conjugated LC compounds such as π-π interaction. An affinity with other LC compounds having fluorinated units (generally inducing dielectric anisotropy) increases by introducing di-fluorinated units on one of the cyclohexane groups of the diluters [17]. Moreover, they are also expected to show high level of photo and thermal stability because the diluters do not have the photo-reactive groups such as the vinyl and the phenyl/phenylene groups [12]. Table 1 lists transition temperatures of R_1-CyFF-Cy-R_2. Decomposition temperatures of most diluters show more than 200 °C, anticipating that these diluters are thermally stable. In particular, the diluters C3-CyFF-Cy-C3 and C2-CyFF-Cy-C5 show the decomposition temperature over 210 °C. The nematic temperature range of C3-CyFF-Cy-C3 is between 28.6 °C and 42.3 °C, whereas that of C2-CyFF-Cy-C5 is between −23.3 °C and 20.0 °C. The result indicates that the nematic temperature of C3-CyFF-Cy-C3 is higher than that of C2-CyFF-Cy-C5. One of assumptions of this result is due to high thermal stability for the C3-CyFF-Cy-C3 since the chemical structure is more symmetrical than that of C2-CyFF-Cy-C5. Different carbon number between two side alkyl groups would increase the steric disorder, leading to relatively unstable nematic phase. Figure 3 indicates DSC thermograms of C3-CyFF-Cy-C3 and C2-CyFF-Cy-C5. The peak observed between crystal (C) and nematic (N) phases of C3-CyFF-Cy-C3 is more obvious than that of C2-CyFF-Cy-C5, probably supporting our assumption. Furthermore, C3-CyFF-Cy-C3 would be more suitable for often-used LCDs, because the LCDs are generally used under the temperature range between 15 °C and 45 °C. Hence, we decided to evaluate the more symmetrical-structured diluter C3-CyFF-Cy-C3 for application of the PS-VA and PI-free IPS LC cells.

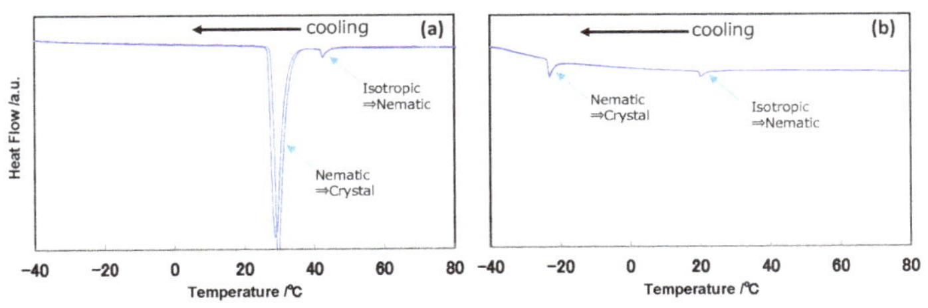

Figure 3. DSC thermograms of (**a**) the C3-CyFF-Cy-C3 and (**b**) the C2-CyFF-Cy-C5.

Table 1. Transition and decomposition temperatures of the diluters R_1-CyFF-Cy-R_2.

Chemical Structure	Product Name	C⇔N Transition Temperature [°C]	N⇔I Transition Temperature [°C]	C⇔I Transition Temperature [°C]	Decomposition Temperature [°C]
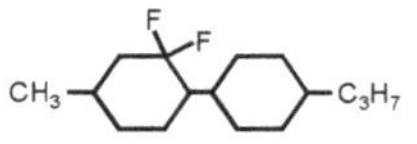	C1-CyFF-Cy-C3	-	-	−5.3	-
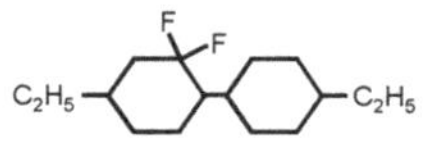	C2-CyFF-Cy-C2	-	-	6.2	-
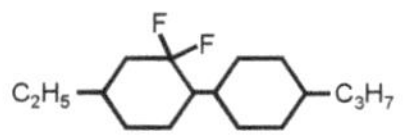	C2-CyFF-Cy-C3	-	-	0.1	201
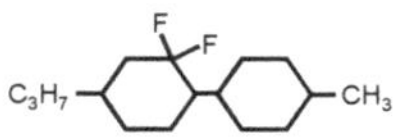	C3-CyFF-Cy-C1	−15.5	5.5	-	210
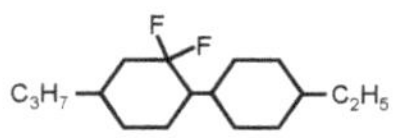	C3-CyFF-Cy-C2	−29.0	2.5	-	206
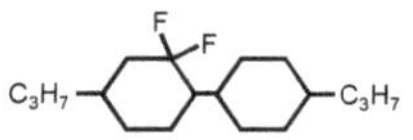	C3-CyFF-Cy-C3	28.6	42.3	-	212
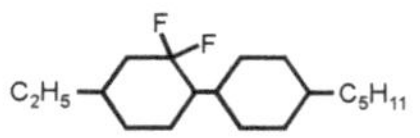	C2-CyFF-Cy-C5	−23.3	20.0	-	212

4. PS-VA Mode

4.1. Materials

In this part, we describe the PS-VA LC cell. For the production of the PS-VA cells, we prepared the LC mixtures, which are consisted of a host LC, the diluter, and the reactive monomer at first. We selected the following two diluters; one is the developed diluter C3-CyFF-Cy-C3 (Figure 4a) in this study, and another is the diluter carrying the vinyl group (C3-HH-V) shown in Figure 4b. In addition, we selected two reactive monomers; one is the reactive monomer carrying the azobenzene moiety (Azo-RM; Figure 5a), and another is the monomer carrying a bi-phenylene moiety (Biph-RM; Figure 5b). The monomer shown in Figure 5b is conventionally used [2]. By combining the host LC, the diluters, and the reactive monomers, three kinds of LC mixtures (LC Mixtures I, II, and III) were prepared, as shown in Table 2.

(a) (b)

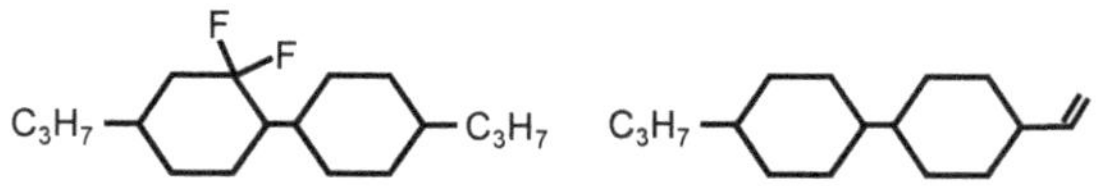

Figure 4. Chemical structures of the diluters (**a**) C3-CyFF-Cy-C3 and (**b**) C3-HH-V.

(a) (b)

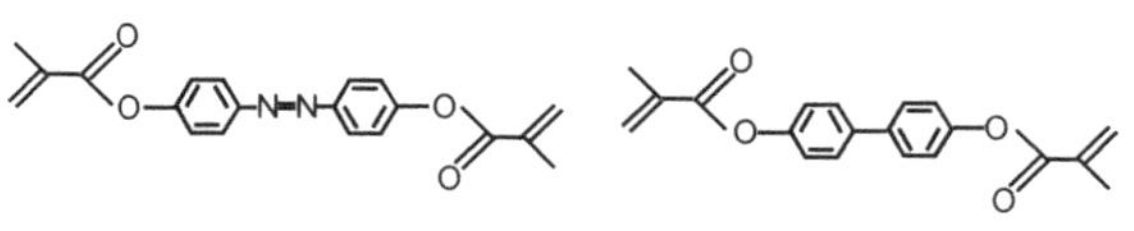

Figure 5. Chemical structures of the reactive monomers including (**a**) azobenzene (Azo-RM) and (**b**) bi-phenylene (Biph-RM) moieties.

Table 2. The LC mixtures I, II, and III including the diluters, host, and reactive monomers for PS-VA LC cells.

	I	II	III	Wt% (LC)	Wt%/LC
Diluter	C3-CyFF-Cy-C3	C30HH-V	C3-CyFF-Cy-C3	25.0	-
RM	Azo-RM	Biph-RM	Biph-RM	-	0.25
Host	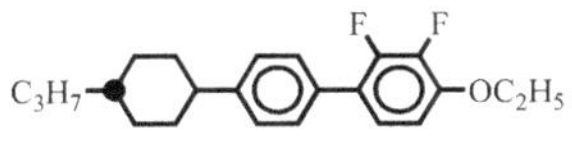			20.0	-
	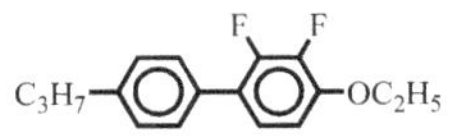			10.0	-
	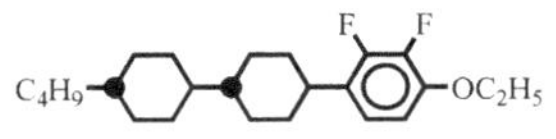			10.0	-
	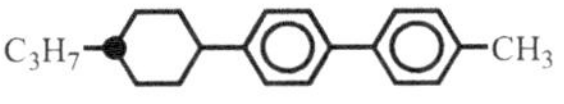			20.0	-
	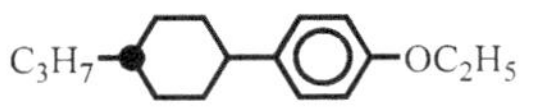			15.0	-

4.2. Preparation of the PS-VA LC Cells

The PS-VA LC cells were prepared as follows. The vertically-aligned PI material was formed on top of ITO substrates, baked at 90 °C for 1 min and 200 °C for 60 min. After baking, photo- and heat-curing sealing material was drawn, and three LC Mixtures, which are shown in Table 2, were each dropped on these substrates. Then, the other substrates, where the same vertically-aligned PI material was formed on them, were laminated on the substrates and heated at 100 °C for 40 min. 10 V square waveform voltage (60 Hz) was applied to each cell, and the black-light (FHF-32BLB developed by Toshiba, in Japan) was exposed to the LC cells at 35 °C for 30 min for preparation of the polymer layers.

4.3. Evaluation of the PS-VA Cell

Response times of PS-VA LC cell were attempted to improve with using the conventional diluter C3-HH-V [10]. However, image sticking has clearly appeared [10,11]. The estimation of the image sticking is the following; the LC mixtures including the C3-HH-V decrease the VHR after exposure of the UV.

Table 3 lists the VHRs of the PS-VA LC cells produced by the LC Mixtures I, II, and III before and after exposure of the black-light. In the LC Mixture I, the VHR was kept high value above 99.2% after exposure of the black-light. In contrast, the VHR decreased significantly for the PS-VA LC cell produced by the LC Mixture II after the black-light exposure. We presume that the result derives from the photo-chemical reaction of the C3-HH-V with the reactive monomer. It was considered that the reactive monomers would generate radicals due to photo-chemical reaction of the monomer with the vinyl group. Hence, for preventing the photo-chemical reaction, we finally designed the symmetrical diluter C3-CyFF-Cy-C3, which does not have the vinyl group and high affinity with the fluorinated LC compounds. As shown in Table 3, the LC Mixture I, which has the diluter C3-CyFF-Cy-C3 and the reactive monomer carrying the azobenzene moiety (Azo-RM), shows high VHR even after the black-light exposure. Therefore, the result indicates that the photo-chemical reaction does not occur between C3-CyFF-Cy-C3 and Azo-RM.

The VHR of the PS-VA LC cell produced by the LC Mixture III, which includes C3-CyFF-Cy-C3 and the conventional-type monomer carrying the biphenylene moiety (Biph-RM), also decreased after the black-light exposure though it was used C3-CyFF-Cy-C3. Since a rate constant of polymerization is relatively slow for the Biph-RM compared with the Azo-RM [18], we presume that unreacted radicals derived from the Biph-RM still exist in the LC layer after the black-light exposure, leading to a significant decrease in the VHR. The results indicate that the LC Mixture I would be the most suitable among the LC Mixtures I, II, and III for the PS-VA LCD.

Table 3. VHRs before and after exposure of the black-light to the PS-VA LC cells.

LC Mixture (PSVA Mode Cell)	VHR (%)	
	Before UV Irradiation	After UV Irradiation
I	99.4	99.2
II	99.3	93.0
III	99.3	97.2

Figure 6 shows the V-T curve of the PS-VA LC cell produced by the LC Mixture I. The threshold voltage Vth indicates about 2.7 V, and the maximum transmittance reaches at 6.4 V. These voltages would be suitable for the real-use. The contrast ratio determined from the ratio of the highest and the lowest transmittance of the V-T curve is over 3000, as listed in Table 4. This is the high level compared with the generally-used LCDs.

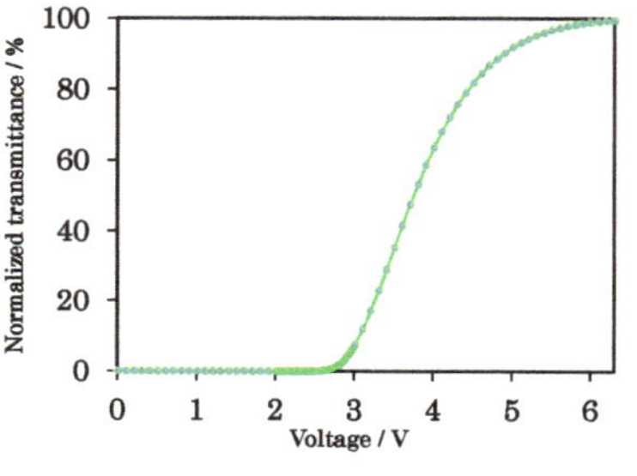

Figure 6. V-T curve for the PS-VA LC cell produced by the LC Mixture I.

Table 4. Contrast ratios and response times of the PS-VA LC cells produced by the LC Mixtures I and II.

	Contrast Ratio	τ_{on} (ms)	τ_{off} (ms)
LC Mixture I	3300:1	9.5	7.3
LC Mixture II	3300:1	9.9	6.7

Response times (τ_{on} and τ_{off}) of the PS-VA LC cell produced by the LC Mixture I were compared with those from the LC Mixture II. Results are also shown in Table 4. Both τ_{on} and τ_{off} are almost equal between the LC Mixtures I and II. The fact supports that the developed diluter C3-CyFF-Cy-C3 would be significantly useful.

Figure 7a,b show the VHRs for the PS-VA LC cells as a function of aging per hour under LED backlight exposure and stored in a 70 °C controlled bath, respectively. Regarding the PS-VA LC cell produced by the LC Mixture I, the VHRs were kept at a high value above 98.0% for 1000 h of the LED backlight exposure and then stored in a 70 °C controlled bath. The result indicates that the photo- and thermal-stability of the LC Mixture I is significantly high. In contrast, the VHRs of PS-VA LC cells produced by both the LC Mixtures II and III decreased under the LED backlight exposure and the storage in 70 °C. The results imply that the photo- and thermal-stability of the LC Mixtures II and III are not enough compared with the LC Mixture I. The VHR of the PS-VA LC cell produced by the LC Mixture II decreased remarkably under the LED backlight exposure. This would be one of the supportive results that the photo-chemical reaction was progressed between the conventional diluter C3-HH-V and the monomer Biph-RM. Thus, we confirm that the combination of the new developed diluter C3-CyFF-Cy-C3 with Azo-RM is significantly useful for the PS-VA LCD, because it shows both fast response speed and high stability.

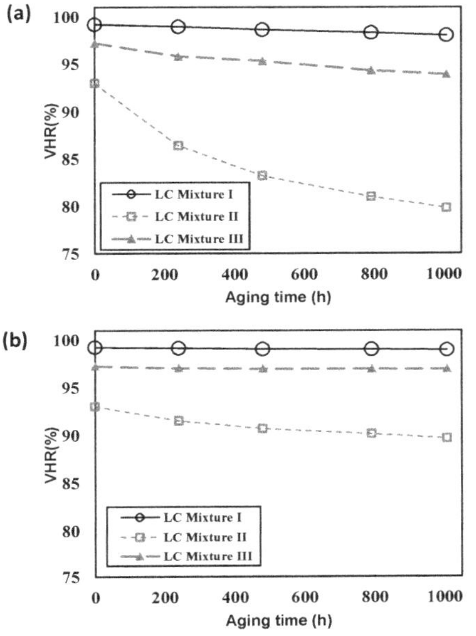

Figure 7. VHR as a function of aging time for PS-VA cells under (**a**) LED backlight exposure, and (**b**) storage in a 70 °C controlled bath.

5. PI-Free IPS Mode

5.1. Materials

In this part, we will describe the PI-free IPS LC cells. For production of the PI-free IPS LC cells, the LC mixtures were firstly prepared. We selected two diluters, C3-CyFF-Cy-C3 and C3-HH-V. The reactive monomer Azo-RM was also selected because the azobenzene unit absorbs polarized UV and induces the uniaxial alignment of the LC [18,19]. For producing the PI-free IPS LC cell, the LC Mixtures IV and V were prepared as shown in Table 5.

5.2. Preparation of the PI-Free IPS Cell

The PI-free IPS LC cells were produced are follows. A pair of IPS mode substrates were prepared; one side of the substrate has pixel and common electrodes, and the other side does not have any electrode. Photo- and heat-curing type sealing material was drawn, and the LC Mixtures IV and V were each dropped on these substrates. The other substrates were then laminated on the substrates and heated at 100 °C for 40 min. The polarized UV was exposed to the LC cell from the normal direction of the LC cell. The exposure of the polarized UV was conducted above nematic-to-isotropic transition temperature (T_{NI}) [18,19]. The homogeneously aligned LC cells were finally obtained.

5.3. Evaluation of the PI-Free IPS Cell

Homogeneously aligned LC was convinced by POM observation, as shown in Figure 8. Before exposure of the polarized UV, the transmittance at axes of 0° and 45° polarized UV to the crossed polarizers are almost same level, and these are not uniform (Figure 8a). This anticipates that the state of the LC alignment was random. After the polarized UV exposure (1 J/cm^2) at 95 °C (above the T_{NI} of the LC composition), dark state was observed at axis of 0° polarized UV to the crossed polarizer, whereas high and uniform transmittance was observed at axis of 45° polarized UV to the crossed polarizers (Figure 8b). The result indicates that the homogeneously aligned LC was obtained by the polarized UV exposure.

Table 5. The LC mixtures IV and V including the diluters, host, and reactive monomers for PI-free IPS LC cells.

	IV	V	Wt% (LC)	Wt%/LC
Diluter	C3-CyFF-Cy-C3	C3-HH-V	25.0	-
RM	Azo-RM	Azo-RM	-	1.0
Host	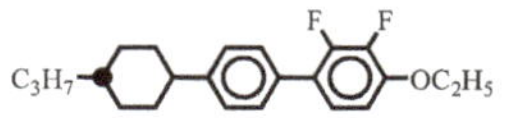		25.0	-
	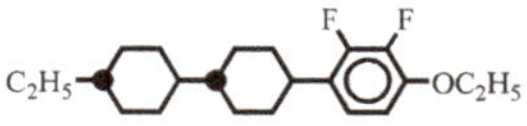		10.0	-
	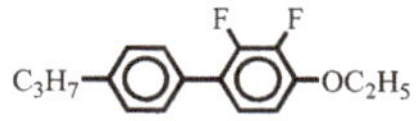		5.0	-
	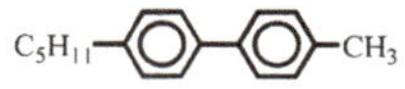		13.0	-
	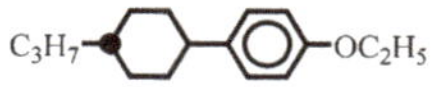		10.0	-
			10.0	-
	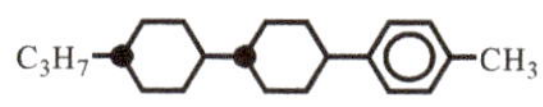		2.0	-

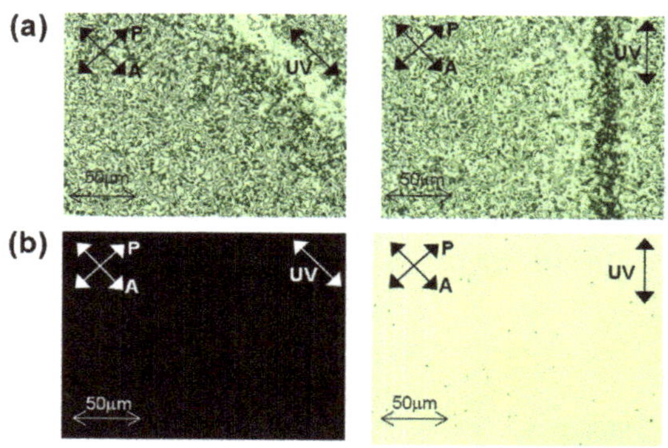

Figure 8. POM images before (**a**) and after (**b**) 1 J/cm^2 polarized UV to the PI-free IPS cells; the angles between crossed polarizers and polarized UV are 0° and 45°.

Table 6 lists the VHR of the PI-free IPS LC cells produced by the LC Mixtures IV and V before and after the polarized UV exposure. Regarding the LC Mixture IV, decrease of the VHR was only slightly after the polarized UV exposure. On the other hand, the VHR of the LC cell produced by the LC Mixture V decreased obviously after exposure of the polarized UV. We presume that the decrease of the VHR is derived from the photo-chemical reaction of the C3-HH-V with Azo-RM. The photo-chemical reaction would induce a generation of radicals, and these radicals would remain in the LC layer of the PI-free IPS LC cell. Therefore, we confirm that using the developed diluter C3-CyFF-Cy-C3 is effective for preventing the photo-chemical reaction with the Azo-RM.

Table 6. VHR before and after exposure of the polarized UV to the PI-free IPS cells.

LC Mixture (PI-Free IPS Cell)	VHR (%)	
	Before UV Irradiation	After UV Irradiation
IV	99.2	98.8
V	99.0	97.3

Figure 9 plots the VHR of the PI-free IPS LC cell produced by the LC Mixture IV as a function of aging hour under LED backlight exposure, and storage in a 70 °C controlled bath, respectively. The VHRs were kept at a high value during 1000 h for both the LED backlight exposure and the storage in the 70 °C atmosphere. In the case of the storage test at 70 °C, the VHR decreased, but only slightly, and the value was approximately 98.0% at 1000 h storage. We estimate that the thermal stability is enough level for real use of the LCD. In the case of the LED backlight exposure, the VHR showed approximately 97.2% after 1000 h aging, indicating that the slight decrease of the VHR was obtained though it is relatively high value compared with the LC mixtures having the conventional diluters. We presume the reason in the following. Since the concentration of the Azo-RM in the Mixture IV is four times higher than that in the Mixture I (Table 2), unreacted monomer molecules and/or polymerized compounds, which are not incorporated into the polymer layer, would be easily remained in the LC layer of the LC Mixture IV. Hence, the residual monomer or polymerized compounds in the LC layer would induce the slight decrease of the VHR, as shown in this figure. Hence, further improvement of the VHR under the LED backlight exposure and the storage in the 70 °C atmosphere is now under consideration.

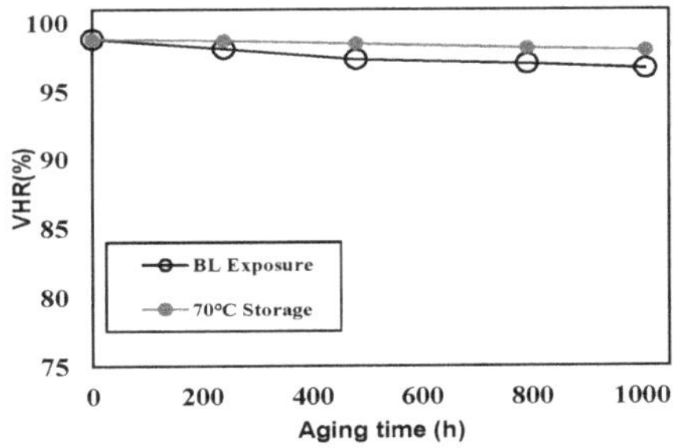

Figure 9. VHR as a function of aging time for the PI-free IPS cells under the LED backlight exposure and the storage in a 70 °C controlled bath.

6. Conclusions

We developed new diluters R_1-CyFF-Cy-R_2 to overcome the tradeoff between the response speed and the image sticking parameter. The newly developed diluters are constructed from two cyclohexane units (Cy) and two side saturated alkyl chains (R_1 and R_2), which are symmetrical in structure. Di-fluorinated groups were introduced into one of the cyclohexane units for increasing the affinity of the other fluorinated LC compounds, which are shown to have negative dielectric anisotropy. The diluters are expected to be significantly stable under the UV because they do not have vinyl groups. In the comparison between two diluters C3-CyFF-Cy-C3 and C2-CyFF-Cy-C5, the temperature showing the nematic phase of C3-CyFF-Cy-C3 is higher than that of C2-CyFF-Cy-C5. This would be due to high thermal stability. We estimate that this is derived from the fact that the chemical structure C3-CyFF-Cy-C3 is more symmetrical than that of C2-CyFF-Cy-C5.

Since the diluter C3-CyFF-Cy-C3 does not have the vinyl groups, the VHRs for both the PS-VA and PI-free IPS LC cells were significantly high after the UV exposure, leading to an extremely small level of image sticking. Moreover, since the diluter C3-CyFF-Cy-C3 carries the di-fluorinated groups on the cyclohexane unit, it shows a high affinity to other fluorinated LC compounds and also shows a fast response speed. We finally succeeded in obtaining the PS-VA LC cell with both the fast response speed and high VHR under both the LED backlight exposure and the storage in the 70 °C controlled bath. We further developed the LC mixture including the Azo-RM for the PI-free IPS LC cell. As the azobenzene moiety absorbs the polarized UV and aligns the LC molecules uniaxially, the homogeneously aligned LC was successfully obtained by the polarized UV exposure to the cell without conventional PI alignment layer. The VHR was kept relatively high after the polarized UV exposure. Therefore, we expect that the high stability PI-free IPS LC cell can be produced by using the diluter C3-CyFF-Cy-C3 and the Azo-RM. As the final conclusion, the combination of the diluter C3-CyFF-Cy-C3 and Azo-RM is significantly useful for the PS-VA and PI-free LCDs.

Author Contributions: Conceptualization, M.M.; Methodology, K.O. and T.S.; Evaluation, M.M.; Simulation, K.O. and T.S.; Writing-original draft preparation, M.M.; Writing-review and editing, M.M., K.O. and T.S. All authors have read and agreed to the published version of the manuscript.

Funding: This research received no external funding.

Institutional Review Board Statement: Not applicable.

Informed Consent Statement: Not applicable.

Data Availability Statement: Not applicable.

Conflicts of Interest: The authors declare no conflict of interest.

References

1. Hanaoka, K.; Nakanishi, Y.; Inoue, Y.; Tanuma, S.; Koike, Y.; Okamoto, K. A new MVA-LCD by polymer sustained alignment technology. In *SID Symposium Digest of Technical Papers*; Blackwell Publishing Ltd.: Oxford, UK, 2004; Volume 35, pp. 1200–1203.
2. Nakanishi, Y.; Hanaoka, K.; Shibasaki, M.; Okamoto, K. Relation between monomer structure and image sticking phenomenon of polymer-sustained-alignment liquid crystal displays. *Jpn. J. Appl. Phys.* **2011**, *50*, 051702. [CrossRef]
3. Lee, S.H.; Kim, S.M.; Wu, S.-T. Emerging vertical-alignment liquid-crystal technology associated with surface modification using UV-curable monomer. *J. Soc. Inf. Disp.* **2009**, *17*, 551–559. [CrossRef]
4. Momoi, Y.; Kwak, M.; Choi, D.; Choi, Y.; Jeong, K.; Koda, T.; Haba, O.; Yonetake, K. Polyimide-free LCD by dissolving dendrimers. *J. Soc. Inf. Disp.* **2012**, *20*, 486–492. [CrossRef]
5. Haba, O.; Itabashi, H.; Sato, S.; Machida, K.; Koda, T.; Yonetake, K.; Kwak, M.; Momoi, Y.; Kim, N.; Hong, S.; et al. UV-induced stable planar alignment of nematic liquid crystals using a polypropyleneimine dendrimer having a mesogen consisting of cinnamate and azobenzene moieties. *Mol. Cryst. Liq. Cryst.* **2015**, *610*, 201–209. [CrossRef]
6. Plummer, E.A.; Klasen-Memmer, M.; Fortte, R.; Haensel, H.; Archetti, G. The evolution of the vertically aligned liquid crystal display. In *SID Symposium Digest of Technical Papers*; Blackwell Publishing Ltd.: Oxford, UK, 2018; Volume 49, pp. 449–452.
7. Lan, S.; Chen, X.; Wei, H.; Wu, L.; Tseng, T.-J.; Hsieh, C.-C.; Chen, H.-H.; Lin, H.-C.; Lee, Y.-H. Self-alignment of liquid crystal for multi-domain liquid-crystal display. In *SID Symposium Digest of Technical Papers*; Blackwell Publishing Ltd.: Oxford, UK, 2018; Volume 49, pp. 453–454.
8. Jeong, I.H.; Yu, J.H.; Jin, H.-S.; Song, K.H.; Kimura, M.; Lee, S.H. Novel approach to achieve conventional polyimide-less IPS/FFS LCDs. In *SID Symposium Digest of Technical Papers*; Blackwell Publishing Ltd.: Oxford, UK, 2014; Volume 45, pp. 1418–1420.
9. Park, Y.; Kim, S.; Lee, E. A study on reducing image-sticking artifacts in wide-screen TFT-LCD monitors. *J. Soc. Inf. Disp.* **2007**, *15*, 969–973. [CrossRef]
10. Saito, M. Liquid Crystal Composition and Liquid Crystal Display Element. Japan Patent 5,565,316 B2, 14 June 2017.
11. Melanie, K.M.; Dagmar, K.; Matthias, B. Liquid-Crystalline Medium. Japan Patent 4,562,357 B2, 13 October 2010.
12. Okamoto, K.; Shibata, T. Cyclohexane Compound and Liquid Crystal Composition Containing the Same. U.S. Patent 8858829 B2, 14 October 2014.
13. Bremer, M.; Naemura, S.; Tarumi, K. Model of ion solvation in liquid crystal cells. *Jpn. J. Appl. Phys.* **1998**, *37*, L88–L90. [CrossRef]
14. Huh, I.-K.; Kim, Y.-B. Fluoro-isothiocyanated liquid crystal materials with high dielectric anisotropy and voltage holding ratio. *Jpn. J. Appl. Phys.* **2002**, *41*, 6466–6470. [CrossRef]
15. Oh-e, M.; Umeda, Y.; Ohta, M.; Aratani, S.; Kondo, K. Unusual voltage-holding ratio characteristics using in-plane switching of nematic liquid crystals. *Jpn. J. Appl. Phys.* **1997**, *36*, L1025–L1028. [CrossRef]
16. Inoue, M. Review of various measurement methodologies of migration ion influence on LCD image quality and new measurement proposal beyond LCD materials. *J. SID* **2020**, *28*, 92–110. [CrossRef]
17. Ichinose, H. Advanced liquid crystal mixture development for high performance liquid crystal displays. *Ekisho* **2011**, *15*, 174–186.

18. Mizusaki, M. Polarised UV-induced homogeneous alignment for fringe-field switching liquid crystal display with a new azobenzene monomer. *Liq. Cryst.* **2019**, *46*, 640–648. [CrossRef]
19. Mizusaki, M.; Tsuchiya, H.; Itoh, T.; Minoura, K. Homogeneous self-alignment technology without forming conventional alignment layers. In *SID Symposium Digest of Technical Papers*; Blackwell Publishing Ltd.: Oxford, UK, 2018; Volume 49, pp. 455–458.

symmetry

MDPI

Article

Flexoelectric Polarization in Liquid Crystalline Elastomers Prepared by Cross-Linking under Horseshoe-Shaped Deformation

Kazuyuki Hiraoka *, Toshio Ishihara, Hiroyuki Minami, Shiori Taira, Katsumi Yamada and Toshihiro Hiejima

Department of Life Science & Sustainable Chemistry, Tokyo Polytechnic University, 1583 Iiyama, Atsugi-shi 243-0297, Japan
* Correspondence: hiraoka@t-kougei.ac.jp

Abstract: Flexoelectric polarization, which is caused by symmetry breaking in a distortion of material, was investigated in liquid crystalline elastomers composed of wedge-shaped mesogens prepared by cross-linking under horseshoe-shaped deformation. X-ray diffractometry suggested that splay distortion along the depth direction was induced in the pseudo-isotropic phase. While almost no electric charge was observed in the smectic A phase, an electric charge caused by polarization due to the flexoelectric effect appeared and reached -1367 pC/mm^2 in the pseudo-isotropic phase. We tentatively conclude that the macroscopic polarization due to the flexoelectric effect emerged and was fixed in the liquid crystalline elastomers by cross-linking under horseshoe-shaped deformation.

Keywords: cross-linked structure; cross section; flexoelectricity; liquid crystalline elastomers; splay; spontaneous polarization

1. Introduction

Liquid crystalline elastomers have been attracting increasing attention as a novel class of liquid crystalline materials because they give rise to new macroscopic features in soft solids with the anisotropic properties of a liquid crystalline phases [1–9]. In particular, they have attracted both scientific and industrial interest as ideal materials for the investigation of piezoelectric and flexoelectric effects, because the polymer network prevents macroscopic flow, which inhibits the emergence of piezoelectricity and flexoelectricity in conventional low-molar-mass liquid crystals. To date, several research groups have demonstrated the mechanical response to electric stimulation attributable to the inverse piezoelectric effect in chiral smectic elastomers, focusing on their potential applications as electrically controllable soft actuators [10–18].

Flexoelectricity, which is associated with the emergence of polarization caused by symmetry breaking by a distortion of material, in liquid crystals as well as liquid crystalline elastomers has been examined in several studies since Meyer predicted its emergence [19–28]. Patel and Meyer [21] and Rudquist et al. [22] pointed out that a periodic splay-bend pattern in the helix of cholesteric liquid crystal couples flexoelectrically to the electric field. Popova et al. reported nonlinear electromechanical coupling due to a local flexoelectric effect in chiral smectic liquid crystals [23]. Meyer et al. [24] predicted and demonstrated an electroclinic effect due to the flexoelectricity in the twist–bend nematic phase. Sreenilayam et al. [25] quantitatively estimated the magnitude of the flexoelectric polarization in bent-core nematic liquid crystals. In addition, flexoelectricity in liquid crystalline elastomers has gained increasing attention since the report of giant flexoelectricity in bent-core nematic liquid crystalline elastomers by Harden et al. [26]. Recently, Hiraoka et al. have reported that the macroscopic polarization due to the flexoelectric effect emerges and is fixed in the liquid crystalline elastomers composed of wedge-shaped mesogens by cross-linking under uniaxial deformation with splay distortion [27]. In addition, they observed an electric field-induced polar deformation resembling the motion of the flukes of a dolphin in the liquid crystalline elastomer [28,29].

Citation: Hiraoka, K.; Ishihara, T.; Minami, H.; Taira, S.; Yamada, K.; Hiejima, T. Flexoelectric Polarization in Liquid Crystalline Elastomers Prepared by Cross-Linking under Horseshoe-Shaped Deformation. *Symmetry* **2023**, *15*, 616. https://doi.org/10.3390/sym15030616

Academic Editor: P. Ulrich Biedermann

Received: 10 January 2023
Revised: 13 February 2023
Accepted: 20 February 2023
Published: 1 March 2023

In this study, the polarization due to the flexoelectric effect is investigated in liquid crystalline elastomers composed of wedge-shaped mesogens prepared by cross-linking under horseshoe-shaped deformation. X-ray measurements of the cross-section of the elastomer film revealed that the director uniformly aligns parallel to the surface in the vicinity of the surface, while it partially inclines 65 degrees to the surface inside the elastomer film. Although no polarization is observed in the smectic A (SmA) phase, it is detected in the pseudo-isotropic phase in which the orientational order partially remains. Here, smectic liquid crystal phases have a layer structure. In the case of the smectic A (SmA) phase, inside the layers the molecules are parallel, on an average, one to each other with their long axes perpendicular to the layer plane [20]. The emergence of polarization suggests that the splay distortion along the depth direction is induced through the transformation from the SmA phase to the pseudo-isotropic phase. The electric charge due to the polarization increases with the temperature in the pseudo-isotropic phase, reaching -1367 pC/mm^2. We confirm that the macroscopic polarization due to the flexoelectric effect emerges and is fixed in the liquid crystalline elastomers by cross-linking under horseshoe-shaped deformation. The liquid crystal elastomers possessing the spontaneous polarization are suitable for their applications as electrically controllable soft actuators.

2. Materials and Methods

An elastomer was synthesized by a hydrosilylation reaction of liquid crystalline side groups with a polysiloxane backbone via a well-known synthesis route [1]. Polymethylhydrosiloxane (polymerization degree of 25–35), undecylenic acid cholesteryl ester, and undecylenic acid 4-undec-10-enoyloxy-phenyl ester were used as the polymer backbone, mesogenic monomer, and cross-linker, respectively. Their chemical structures are shown in Figure 1a. It is noteworthy that the cholesterol-derived mesogen seems to be a wedge-shaped molecule (Figure 1b). As previously reported, the electric dipole moment p_i along the molecular long axis of the cholesterol-derived mesogen was estimated to be about 1.1 D by calculation after the conformational structure of the mesogen was optimized by the semiempirical molecular orbital method [27,29,30]. Wedge-shaped molecules carrying permanent dipole moments, such as the cholesterol-derived mesogens used here, yield macroscopic polarization $P \neq 0$ when splay distortion is imposed [19,20].

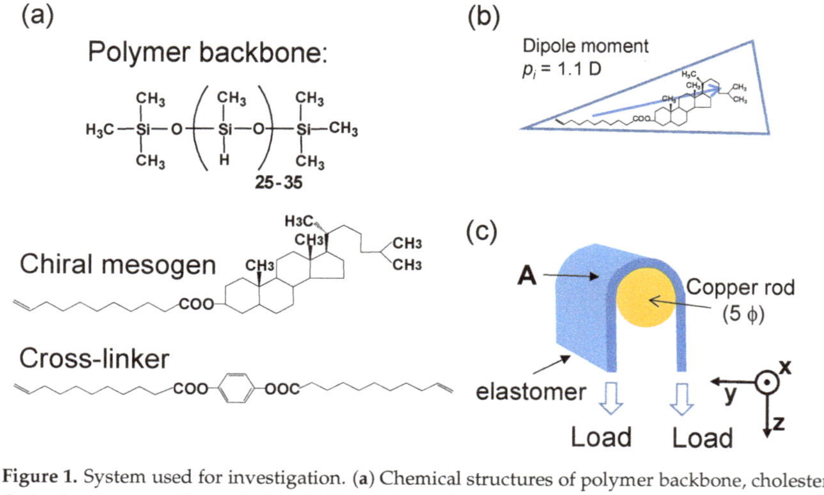

Figure 1. System used for investigation. (**a**) Chemical structures of polymer backbone, cholesterol-derived mesogen, and cross-linker. (**b**) The cholesterol-derived mesogen is a wedge-shaped molecule. The arrow indicates the direction of electric dipole moment. (**c**) Sample preparation of a horseshoe-like deformed liquid crystalline elastomer.

The elastomer exhibited the following phase sequence [glassy state 25 °C smectic A phase 110 °C pseudo-isotropic phase] upon heating. As already reported, these transition temperatures were determined by differential scanning calorimetry (DSC), thermomechanical analysis (TMA), and temperature-dependent X-ray diffractometry [27–29]. Because X-ray diffractometry has revealed that the orientational order partially remained at about $S = 0.2$, even in the temperature region of the isotropic phase of the liquid crystalline elastomers, as described in previous papers, the temperature region of the isotropic phase here is designated as the pseudo-isotropic phase in this report [27–29,31,32].

The elastomer film was prepared by a spin-casting technique in a solution containing the polysiloxane (2 mmol), the mesogen (1.6 mmol), the cross-linker (0.2 mmol), and a Pt catalyst [platinum (0)-1,3-divinyl-1,1,3,3-tetramethyl-disiloxane complex, Aldrich] in 2 mL of toluene. The reaction was performed under centrifugation (6000 rpm) at 40 °C for 4 h. Thereafter, the reaction vessel was cooled to room temperature and the elastomer, which was swollen with toluene, was carefully removed from the vessel. To obtain a splay distortion of the director, the swollen elastomer is deformed into a horseshoe shape by loading it with a stress of 89.5 mN/mm^2 for 24 h at room temperature, as shown in Figure 1c, in which the loading direction is indicated by a pair of arrows. During the deformation process at room temperature (25 °C), the toluene evaporated continuously from the network, and successive phase transformations occurred from the isotropic phase of the gel to the liquid crystalline phase of the dry network during the sample preparation. Since the cross-linking reaction simultaneously proceeds in the deformation process, the anisotropic orientation of the director in the horseshoe-shaped deformation is chemically locked in the polymer network.

The electric charge was measured using a charge meter (Kistler, Charge meter type 5015) by contacting the tip of a hand-made electrode assembled with a crocodile clip to the surface of a sample at the position marked by arrow A in Figure 1c. The contact area was estimated to be about 1.5 mm^2. The sample was placed on a temperature-controlled hot plate.

The molecular alignment and orientation in the horseshoe-shaped part of the deformed elastomer [arrow A in Figure 1c] were confirmed by X-ray diffractometry (Rigaku, Nanowier + Micro07HFM using a Cu-K$_\alpha$ beam filtered by a confocal mirror. X-ray wavelength = 1.54 Å, X-ray power = 2.7 kW). The X-ray beam was narrowed by passing it through a slit (0.4 mm ϕ). The X-ray measurements were performed using a two-dimensional hybrid pixel array detector (Rigaku Oxford Diffraction, PILATUS200K). The X-rays scattered on the y–z plane were measured after maintaining the measurement temperature for at least 10 min. The sample was mounted in a microfurnace (Mettler, FP82HT) during the X-ray measurements.

3. Results and Discussion

X-ray diffractometry was performed at seven positions to investigate the cross-sectional profile of the macroscopic orientation of the director in the horseshoe-like deformed part of the elastomer at the arrow A position in Figure 1c. Figure 2a shows a photograph of the cross-section of the sample. The observed seven positions are designated consecutively from position 1 to position 7, as illustrated in Figure 2b. Position 1 is nearest the outer surface and position 7 is nearest the inner surface. Position 4 is located near the center of the depth profile. The places of position 5 and the position 6 are shown in Figure 2b. Here, the interval between positions is 0.25 mm and the sample thickness is about 1.5 mm.

Figure 3 shows the results of a series of X-ray analyses at these seven positions. The X-ray measurements were performed at room temperature. Figure 3(1-a)–(1-c) show the X-ray diffraction pattern, the azimuthal X-ray intensity profile, and the molecular orientational model based on the X-ray results at position 1, respectively. As similarly reported in previous papers on uniaxially deformed elastomers [27–29], the orthogonal alignment due to the SmA structure is confirmed by the X-ray diffraction pattern in Figure 3(1-a), in which the small-angle reflections due to the smectic layers and the wide-angle diffractions associated with the molecular arrangement of the mesogenic side groups within the smectic layers exist. From the analysis of the azimuthal intensity profile of the wide-angle diffraction shown in Figure 3(1-b), we estimate the macroscopic order parameter as $S = (1/2) < 3\cos^2\alpha - 1 > = 0.83$ [33]. The two characteristic reflections indicate that the mesogens and smectic layers are macroscopically uniformly aligned near the surface at position 1, as illustrated in Figure 3(1-c). Compared with the cross-sectional image in Figure 2b, the direction of the director in Figure 3(1-c) seems to be almost parallel to the sample surface.

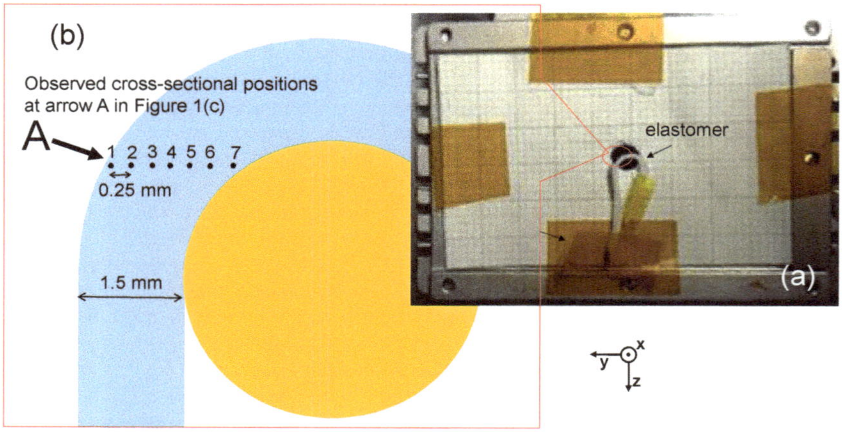

Figure 2. (**a**) Photograph of cross-section of the sample. (**b**) The seven positions observed by X-ray diffraction are designated consecutively from position 1 to position 7 as schematically shown.

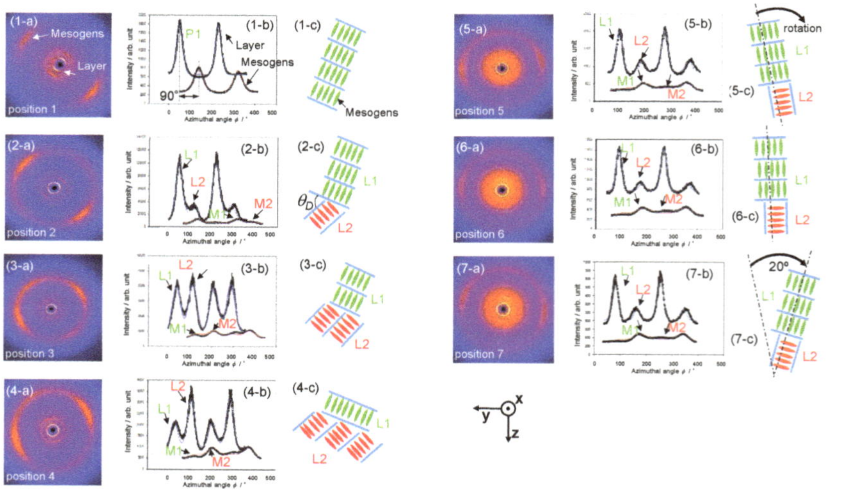

Figure 3. X-ray analysis of seven positions in cross-section of horseshoe-shaped part of deformed liquid crystalline elastomer. X-ray diffraction patterns are shown in (**1-a**)–(**7-a**), azimuthal X-ray intensity profiles are shown in (**1-b**)–(**7-b**), and molecular orientational models based on X-ray results are shown in (**1-c**)–(**7-c**).

Figure 3(2-a)–(2-c) show the X-ray diffraction pattern, the azimuthal intensity profile, and the molecular orientational model at position 2, respectively. The small-angle reflections due to smectic layers and the wide-angle diffractions associated with the mesogenic side groups are also recognized in Figure 3(2-a),(2-b). It is noteworthy that two peaks are observed in the azimuthal profile of layer reflection at a small angle (L1 and L2) and mesogenic diffraction at a wide angle (M1 and M2) in Figure 3(2-b). The azimuthal profile with two peaks indicates that two smectic domains exist. From the intensity ratio of peak L1 to peak L2 in Figure 3(2-b), the volume ratio of two smectic domains can be estimated to be 3:1, as illustrated in Figure 3(2-c). The X-ray analysis of the larger domain due to peak L1 indicates that the direction of the director seems to be almost parallel to the sample surface, the same as the director at position 1, while the director of another smaller domain due to peak L2 inclines at about q_D degrees to the director of the larger domain parallel to the surface. In addition, angle q_D between the two domains is estimated to be about 65 degrees.

Figure 3(3-a)–(3-c) show the X-ray pattern, the azimuthal profile, and the molecular orientational model at position 3, respectively. Two peaks are also observed in the azimuthal profile of layer reflection (L1 and L2) and mesogenic diffraction (M1 and M2) in Figure 3(3-b). The intensity of peak L1 is almost the same as that of peak L2 in the azimuthal profile in Figure 3(2-b). This means that the volume ratio of two smectic domains can be estimated as 1:1 (=2:2) at position 3, as illustrated in Figure 3(3-c).

Figure 3(4-a)–(4-c) show the X-ray results at position 4 near the center of the depth profile. Two peaks are also observed in the azimuthal profile of layer reflection (L1 and L2) and mesogenic diffraction (M1 and M2) in Figure 3(4-b). The intensity of peak L2 becomes higher than that of peak L1 at position 4. From the peak intensity ratio between peak L1 and peak L2 in Figure 3(4-b), the volume ratio of two smectic domains is estimated to be 1:3, as illustrated in Figure 3(4-c).

Figure 3(7-a)–(7-c) show the X-ray results at position 7 near the inner surface. Two peaks are also observed in the azimuthal profile of layer reflection at a small angle (arrows L1 and L2) and mesogenic diffraction at a wide angle (M1 and M2) in Figure 3(7-b). From the peak intensity ratio between peak L1 and peak L2 in Figure 3(7-b), the volume ratio of two smectic domains is estimated to be 1:3, as illustrated in Figure 3(7-c). In a similar manner to the case of position 2 at the second nearest position from the outer surface, the direction of the larger domain due to L1 is found to be almost parallel to the surface of the inside diameter, as illustrated in Figure 3(7-c).

The X-ray results at positions 5 and 6 are shown in Figure 3(5-a)–(5-c),(6-a)–(6-c), respectively. While the azimuthal intensity profiles at positions 5 and 6 in Figure 3(5-b),(6-b), respectively, resemble that at position 7 in Figure 3(7-b), the direction of the X-ray pattern rotates about 20 degrees clockwise from position 5 [Figure 3(5-a)] to position 7 [Figure 3(7-a)]. The molecular orientational models based on the X-ray results at positions 5, 6, and 7 are illustrated in Figure 3(5-c),(6-c),(7-c), respectively.

The molecular alignment models shown in Figure 3(1-c)–(7-c) are combined in Figure 4a to illustrate the cross-section of the elastomer film and discuss the depth profile of molecular alignments. As mentioned above, molecules align uniformly and the direction of the director is almost parallel to the surface in the area of the outer surface at position 1 in Figure 4a. Here, the smectic domain, where the director is parallel to the surface, is tentatively designated as "the parallel domain". Another domain, where the director inclines at 65 degrees to the director of the parallel domain, appears at position 2, which is a depth of 0.25 mm from the surface. This new domain is designated as "the inclined domain". The peak intensity ratio between the inclined domain and the parallel domain, which is 1:4 at position 2 (0.25 mm depth), increases as the observed position becomes deeper. It becomes 2:2 (1:1) at position 3 (0.50 mm depth) and then 3:1 at position 4 (0.75 mm depth). In other words, the inclined domain, which does not exist at the surface, becomes larger than the parallel domain at the center position in the cross-sectional profile (0.75 mm depth). In addition, the direction of the domain rotates about 20 degrees clockwise from position 5 to position 7. The director of the larger domain, indicated by L1, at position 7 is almost parallel to the inner surface.

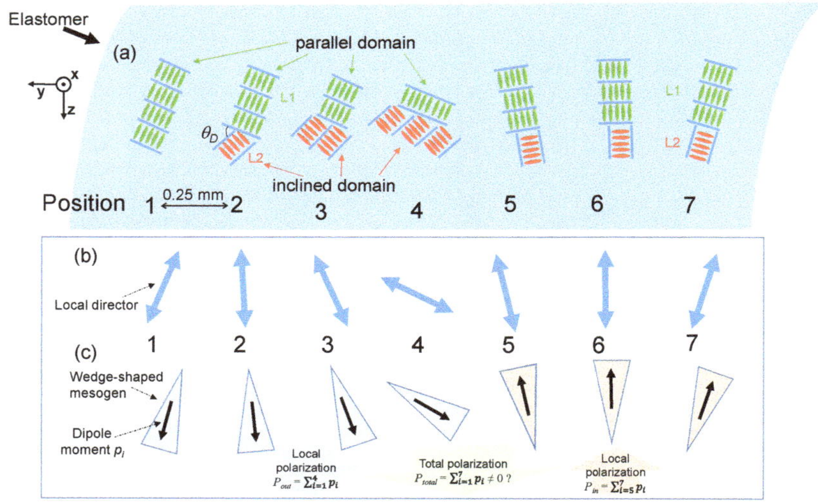

Figure 4. Molecular orientation model based on X-ray analysis of the horseshoe-like deformed liquid crystalline elastomer. (**a**) Schematic model of molecular orientation in cross-section of the liquid crystalline elastomer based on results of X-ray analysis. (**b**) Hypothetical model of the direction of director in the pseudo-isotropic phase. (**c**) The flexoelectric polarization due to splay distortion may be locally induced in the pseudo-isotropic phase.

Figure 4b shows a hypothetical model of the direction of local directors in the cross-section of the elastomer in the pseudo-isotropic phase where the orientational order partially remains. The direction of the local director at each position is thought to be governed by the volume ratio of the parallel domain to the inclined domain in the SmA phase. The director is uniformly parallel to the surface at position 1. The director at position 4 is assumed to incline at about 48 degrees (=65° × 3/4) to the director of position 1, as illustrated in Figure 4b, because three-fourths of the area at position 4 corresponds to the domain which inclines 65 degrees relative to the surface. The change in the direction of the local director from position 1 to position 4 seems to induce splay distortion in the pseudo-isotropic phase. Here, splay distortion is defined as a distortion of the director $\boldsymbol{n}$ expressed by (div $\boldsymbol{n}$)^2 [20]. Because the mesogens used here are wedge-shaped, as shown in Figure 1b, the macroscopic polarization $P_{out} = \sum_{i=1}^{4} p_i$ caused by the flexoelectric effect can simultaneously emerge in the horseshoe-like deformed elastomer because the splay distortion is fixed by cross-links, as illustrated in positions 1–4 in Figure 4c [19,20,28].

Subsequently, let us discuss the change in the direction of the local director from position 5 to position 7 shown in Figure 4b. There is almost no change in the ratio of the inclined domain to the parallel domain among positions 5–7, while the direction of the local domain rotates about 20 degrees clockwise from position 5 to position 7. The rotation of the director induces another splay distortion in the pseudo-isotropic phase. The flexoelectric polarization $P_{in} = \sum_{i=5}^{7} p_i$ due to the splay distortion also emerges in the region from position 5 to position 7, as illustrated in Figure 4c.

As the total polarization P_{total} ($P_{total} = \sum_{i=1}^{7} p_i = \sum_{i=1}^{4} p_i + \sum_{i=5}^{7} p_i \neq 0$) is apparently not canceled out, the macroscopic polarization caused by the flexoelectric effect consequently emerges in the horseshoe-like deformed elastomer in which splay distortion is fixed by cross-links (Figure 4c). The temperature dependence of the surface charge density is measured to confirm the emergence of the polarization in the liquid crystalline elastomer. Two samples are used for investigation: silicone rubber with a thickness of 1.0 mm as the reference sample and the horseshoe-shaped part of the deformed elastomer with splay distortion. Figure 5 shows the temperature dependence of the surface charge density of these two samples. No electric charge is observed in the temperature region measured between 20 °C and 160 °C in the silicone rubber. Electric charge is also not observed below 100 °C in the horseshoe-shaped part of the deformed elastomer, because the layered structure of the SmA phase disturbs the emergence of splay distortion of the director. An electric charge appears and is estimated to be −410 pC/mm^2 at 110 °C in the vicinity of the transition temperature from the SmA phase to the pseudo-isotropic phase. The macroscopic polarization probably emerges because of splay distortion in the pseudo-isotropic phase with the partial orientational order of mesogens.

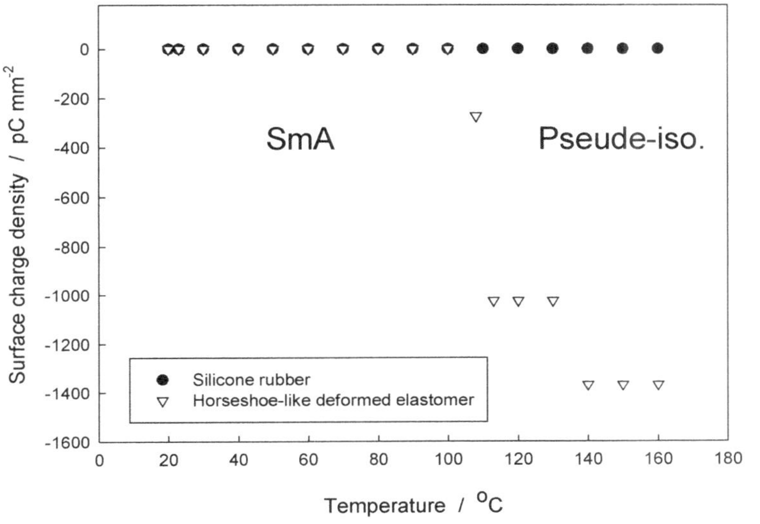

Figure 5. Temperature dependence of surface charge density of two samples: silicone rubber used as a reference sample and the horseshoe-shaped part of deformed elastomer.

Moreover, the surface charge density increases at a temperature above 110 °C. It reaches −1367 pC/mm^2 above 140 °C in the temperature region of the pseudo-isotropic phase. In other words, the surface charge density starts to increase in the vicinity of the transition temperature from the SmA phase to the pseudo-isotropic phase, and it reaches −1367 pC/mm^2 in the pseudo-isotropic phase in which the splay distortion of the director, due to the remaining orientational order of mesogens, is allowed. As reported in a previous paper [27], the uniaxially deformed liquid crystalline elastomer with uniform orientation of the same mesogens and almost no and/or a small electric charge was observed in the same measured temperature region including the SmA phase as well as the pseudo-isotropic phase, because neither splay nor bend distortion occurred in the uniformly oriented sample. In other words, macroscopic polarization hardly appears in the uniformly aligned liquid crystalline elastomer, in which the dipole moment is macroscopically cancelled out in the symmetric alignment. It may be safely concluded that the splay distortion of wedge-shaped mesogens with a permanent dipole moment is fixed by cross-links in the pseudo-isotropic phase with nematic order and brings about the macroscopic polarization due to flexoelectricity, and that the observed electric charge is caused by the polarization.

4. Conclusions

A liquid crystalline elastomer with wedge-shaped mesogens derived from cholesterol was synthesized with horseshoe-shaped deformation. X-ray diffractometry revealed a cross-sectional profile of the director in the horseshoe-like deformed part of the elastomer in the SmA phase; the director uniformly aligned parallel to the surface in the outer surface area, while it was partially inclined about 65 degrees to the surface inside the elastomer film. Although no polarization was observed in the SmA phase, the polarization was detected in the pseudo-isotropic phase of the horseshoe-like deformed part of the elastomer. The emergence of polarization suggested that the splay distortion along the depth direction is induced through the transformation from the SmA phase to the pseudo-isotropic phase. The surface charge density due to the polarization increased with the temperature in the pseudo-isotropic phase and reached $-1367\ \mathrm{pC/mm^2}$. It was concluded that the macroscopic polarization due to the flexoelectric effect emerged and was fixed in the elastomer within the splay distortion cross-linked in the horseshoe-like deformed liquid crystalline elastomer.

Author Contributions: Formal analysis, T.I. and H.M.; Investigation, K.H. and S.T.; Supervision, K.Y. and T.H. All authors have read and agreed to the published version of the manuscript.

Funding: This research was supported by a Grant-in-Aid for Scientific Research (C) (#20K05654) from the Ministry of Education, Culture, Sports, Science and Technology (MEXT).

Institutional Review Board Statement: Not applicable.

Informed Consent Statement: Not applicable.

Data Availability Statement: Not applicable.

Acknowledgments: We would like to thank Syunta Asada and Naoki Shimura for valuable discussions.

Conflicts of Interest: The authors declare no conflict of interest.

References

1. Finkelmann, H.; Kock, H.J.; Rehage, G. Liquid crystalline elastomers—A new type of liquid crystalline material. *Makromol. Chem. Rapid Commun.* **1981**, *2*, 317–322. [CrossRef]
2. Zentel, R. Liquid crystalline elastomers. *Angew. Chem. Adv. Mater.* **1989**, *101*, 1437–1445. [CrossRef]
3. Brand, H.R.; Finkelmann, H. Physical properties of liquid crystalline elastomers. In *Handbook of Liquid Crystals*; Demus, D., Goodby, J., Gray, G.W., Spiess, H.W., Vill, V., Eds.; Wiley-VCH: Weinheim, Germany, 1989; Volume 3, pp. 277–302.
4. Warner, M.; Terentjev, E.M. *Liquid Crystal Elastomers*; Revised Edition; Clarendon Press: Oxford, UK, 2007; pp. 1–46.
5. De Gennes, P.G. Un muscle artificiel semi-rapide. *C. R. Acad. Sci.* **1997**, *324*, 343–348. [CrossRef]
6. Neufeld, R.A.E.; Shahsavan, H.; Zhao, B.X.; Abukhdeir, N.M. Simulation-based design of thermally-driven actuators using liquid crystal elastomers. *Liq. Cryst.* **2018**, *45*, 1010–1022. [CrossRef]
7. Skačej, G. Elastocaloric effect in liquid crystal elastomers from molecular simulations. *Liq. Cryst.* **2018**, *45*, 1964–1969. [CrossRef]
8. Ube, T.; Yoda, T.; Ikeda, T. Fabrication of photomobile polymer materials with phase-separated structure of crosslinked azobenzene liquid-crystalline polymer and poly(dimethylsiloxane). *Liq. Cryst.* **2018**, *45*, 2269–2273. [CrossRef]
9. Braun, L.B.; Zentel, R. Functional liquid crystalline particles and beyond. *Liq. Cryst.* **2019**, *46*, 2023–2041. [CrossRef]
10. Lehmann, W.; Skulpin, H.; Tolksdorf, C.; Gebhard, E.; Zentel, R.; Krüger, P.; Lösche, M.; Kremer, F. Giant lateral electrostriction in ferroelectric liquid-crystalline elastomers. *Nature* **2001**, *410*, 447–450. [CrossRef]
11. Köhler, R.; Stannarius, R.; Tolksdorf, C.; Zentel, R. Electroclinic effect in free-standing smectic elastomer films. *Appl. Phys. A* **2005**, *80*, 381–388. [CrossRef]
12. Spillmann, C.M.; Ratna, B.R.; Naciri, J. Anisotropic actuation in electroclinic liquid crystal elastomers. *Appl. Phys. Lett.* **2007**, *90*, 021911. [CrossRef]
13. Spillmann, C.M.; Kapur, A.V.; Bentrem, F.W.; Naciri, J.; Ratna, B.R. Critical field strength in an electroclinic liquid crystal elastomer. *Phys. Rev. Lett.* **2010**, *104*, 227802. [CrossRef]
14. Eckert, T.; Finkelmann, H. Piezoelectricity of mechanically oriented Sc*-elastomers. *Macromol. Rapid Commun.* **1996**, *17*, 767–773. [CrossRef]
15. Adams, J.M.; Warner, M.; Stenull, O.; Lubensky, T.C. Smectic-A elastomers with weak director anchoring. *Phys. Rev. E* **2008**, *78*, 011703. [CrossRef] [PubMed]
16. Hiraoka, K.; Stein, P.; Finkelmann, H. Electromechanics of a chiral smectic C elastomer: Measurement of complex piezoelectric constant through successive phase transformation. *Macromol. Chem. Phys.* **2004**, *205*, 48–54. [CrossRef]
17. Hiraoka, K.; Kobayashi, M.; Kazama, R.; Finkelmann, H. Electromechanics of monodomain chiral smectic C elastomer: Mechanical response to electric stimulation. *Macromolecules* **2009**, *42*, 5600–5604. [CrossRef]
18. Hiraoka, K.; Kishimoto, T.; Kato, M.; Tashiro, T. Electroclinic and electromechanical effects of a side-chain chiral smectic A elastomer. *Liq. Cryst.* **2011**, *38*, 489–493. [CrossRef]
19. Meyer, R.B. Piezoelectric effects in liquid crystals. *Phys. Rev. Lett.* **1969**, *22*, 918–921. [CrossRef]
20. De Gennes, P.G.; Prost, J. *The Physics of Liquid Crystals*, 2nd ed.; Oxford Science Publications: Oxford, UK, 1993; pp. 10–20, 135–139.
21. Patel, J.S.; Meyer, R.B. Flexoelectric electro-optics of a cholesteric liquid crystal. *Phys. Rev. Lett.* **1987**, *58*, 1538–1540. [CrossRef]

22. Rudquist, P.; Komitov, L.; Lagerwall, S.T. Linear electro-optic effect in a cholesteric liquid crystal. *Phys. Rev. E* **1994**, *50*, 4735. [CrossRef]
23. Popova, E.V.; Kopeychenko, E.I.; Krivoshey, V.V.; Fedoryako, A.P. Piezoelectric and flexoelectric effects in ferroelectric liquid crystals. *Phys. Rev. E* **2012**, *86*, 031705. [CrossRef]
24. Meyer, C.; Luckhurst, G.R.; Dozov, I. Flexoelectrically driven electroclinic effect in the twist-bend nematic phase of achiral molecules with bent shapes. *Phys. Rev. Lett.* **2013**, *111*, 067801. [CrossRef] [PubMed]
25. Sreenilayam, S.; Panarin, Y.; Vij, J.K.; Torgova, S.I.; Lehmann, A.; Tchierske, C. Flexoelectric polarization studies in bent-core nematic liquid crystals. *Phys. Rev. E* **2015**, *92*, 022502. [CrossRef] [PubMed]
26. Harden, J.; Chambers, M.; Verduzco, R.; Luchette, P.; Gleeson, J.T.; Sprunt, S.; Jakli, A. Giant flexoelectricity in bent-core nematic liquid crystal elastomers. *Appl. Phys. Lett.* **2010**, *96*, 102907. [CrossRef]
27. Hiraoka, K.; Ishihara, T.; Minami, H.; Taira, S.; Komesu, S.; Yamada, K.; Oshima, M. Spontaneous polarization due to flexoelectric effect in liquid crystalline elastomers prepared by cross-linking under splay distortion. *Liq. Cryst.* **2022**, *49*, 2051–2057. [CrossRef]
28. Hiraoka, K.; Kashimura, H.; Tanaka, S.; Kaneshima, S. Electric-field-induced deformation of chiral smectic A liquid-crystalline elastomers composed of cholesterol derivative mesogens. *Mol. Cryst. Liq. Cryst.* **2017**, *646*, 168–175. [CrossRef]
29. Hiraoka, K.; Taira, S.; Hoshino, Y.; Ishihara, T.; Yamada, K.; Oshima, M. Electric-field-induced deformation caused by electroclinic and flexoelectric effects in liquid-crystalline elastomer with wedge-shaped mesogens derived from cholesterol. *Liq. Cryst.* **2021**, *47*, 1535–1545. [CrossRef]
30. Frisch, M.J.; Trucks, G.W.; Schlegel, H.B.; Scuseria, G.E.; Robb, M.A.; Cheeseman, J.R.; Scalmani, G.; Barone, V.; Petersson, G.A.; Nakatsuji, H.; et al. *Gaussian 16, Revision B.01*; Gaussian, Inc.: Wallingford, CT, USA, 2016.
31. Hiraoka, K.; Tagawa, N.; Baba, K. Shape-Memory effect controlled by the crosslinking topology in uniaxially-deformed smectic C^* elastomers. *Macromol. Chem. Phys.* **2008**, *209*, 298–307. [CrossRef]
32. Rössle, M.; Braun, L.; Schollmeyer, D.; Zentel, R.; Lagerwall, J.P.F.; Giesselmann, F.; Stannarius, R. Differences between smectic homo- and co-polysiloxanes as a consequence of microphase separation. *Liq. Cryst.* **2005**, *32*, 533–538. [CrossRef]
33. Mitchell, G.R.; Windle, A.H. Orientation in liquid crystal polymers. In *Developments in Crystalline Polymers-2*; Bassett, D.C., Ed.; Elsevier Applied Science: Essex, UK, 1988; Chapter 3.

symmetry

MDPI

Review

Blue Phase Liquid Crystals with Tailored Crystal Orientation for Photonic Applications

SeongYong Cho and Masanori Ozaki *

Division of Electrical, Electronic and Infocommunication Engineering, Graduated School of Engineering, Osaka University, 2-1 Yamada-Oka, Suita, Osaka 565-0871, Japan; scho@opal.eei.eng.osaka-u.ac.jp
* Correspondence: ozaki@eei.eng.osaka-u.ac.jp

Abstract: Blue phase (BP) liquid crystals, which self-assemble into soft three-dimensional (3D) photonic crystals, have attracted enormous research interest due to their ability to control light and potential photonic applications. BPs have long been known as optically isotropic materials, but recent works have revealed that achieving on-demand 3D orientation of BP crystals is necessary to obtain improved electro-optical performance and tailored optical characteristics. Various approaches have been proposed to precisely manipulate the crystal orientation of BPs on a substrate, through the assistance of external stimuli and directing self-assembly processes. Here, we discuss the various orientation-controlling technologies of BP crystals, with their mechanisms, advantages, drawbacks, and promising applications. This review first focuses on technologies to achieve the uniform crystal plane orientation of BPs on a substrate. Further, we review a strategy to control the azimuthal orientation of BPs along predesigned directions with a uniform crystal plane, allowing the 3D orientation to be uniquely defined on a substrate. The potential applications such as volume holograms are also discussed with their operation principle. This review provides significant advances in 3D photonic crystals and gives a huge potential for intelligent photonic devices with tailored optical characteristics.

Keywords: blue phase liquid crystals; photonic crystals; 3D orientation control; photonic devices

Citation: Cho, S.; Ozaki, M. Blue Phase Liquid Crystals with Tailored Crystal Orientation for Photonic Applications. *Symmetry* **2021**, *13*, 1584. https://doi.org/10.3390/sym13091584

Academic Editors: Shoichi Ishihara and Sadahito Uto

Received: 21 July 2021
Accepted: 23 August 2021
Published: 27 August 2021

1. Introduction

Blue phase (BP) liquid crystals (LCs), which are a kind of chiral LC state, have emerged significant research interest in a variety of photonic applications owing to their three-dimensional (3D) photonic structures. They are built by self-assembling so-called double twist cylinders (DTCs) into two types of cubic symmetry: body-centered cubic (BP I, $I4_132$) and simple cubic (BP II, $P4_232$), depending on the chiral twisting power and temperature (Figure 1) [1–3]. The highly ordered structures of BPs give rise to a unique selective Bragg reflection in the visible light region, which only reflect a circular-polarized (CP) light with the same handedness as the helix [4,5]. The central wavelength of the selective Bragg reflection (λ_c) is expressed as

$$\lambda_c = \frac{2na}{\sqrt{h^2 + k^2 + l^2}} \cos\theta \tag{1}$$

where n is the average refractive index of BPs, a is the lattice constant of BPs, (hkl) is the Miller indices representing the crystal orientation, and θ is the propagation angle of light in BPs with respect to the normal direction of a Miller plane. Because of the external stimuli-responsivity of BPs, they can be easily switched and tuned by controlling temperature [6,7], by applying an electric field [7–11], and by illuminating light [12,13], making them applicable to various photonic applications such as displays [14], photonic crystal lasers [6,15–18], biosensors [19,20], optical filters [21], polarization converters [22], and diffractive optics with the capability of nonmechanical beam steering [23,24].

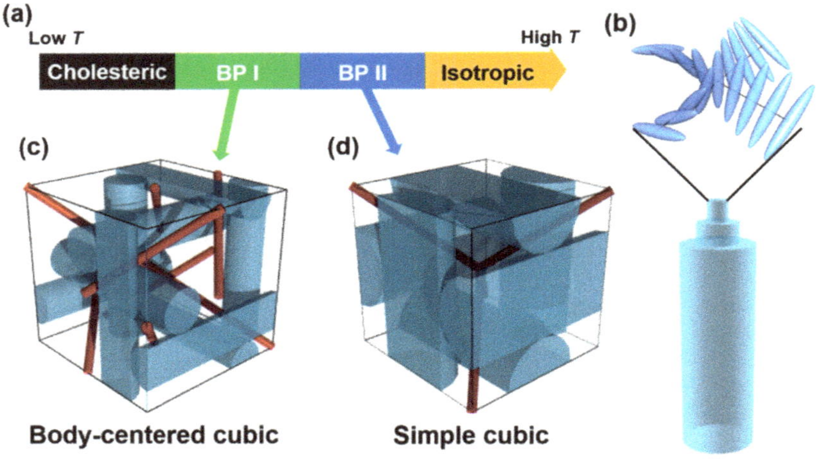

Figure 1. (**a**) Temperature range in which BP I and BP II appear. (**b**) Structure of a double-twist cylinder (DTC) composed of biaxial twist of the LC molecules. (**c**,**d**) Unit structures of body-centered cubic BP I (**c**) and simple cubic BP II (**d**), with their dislocation lines (red lines).

Despite the technological potential of BPs, several issues have been raised to achieve tailored light-matter interaction for practical applications. First, BPs typically appear in a narrow temperature range of approximately 1–2 °C between cholesteric (Ch) and isotropic state. BPs that exhibit over a wide range of temperature were obtained by adding a small amount of photo-reactive monomers and in situ polymerization, providing a temperature range of BPs over 50 °C including room temperature [25,26]. Other strategies have also been proposed to extend the temperature range of BPs, by dispersing nanoparticles [27,28], bent-core molecules [6,29], and T-shaped molecules [30], by confining BPs in the microsized walls [31–33], and by designing the chemical structure of LC molecules [34]. Second, cubic BPs generally exhibit polycrystalline structures, which are composed of small platelet domains with different crystallographic orientations. Because of the complicated structures of BPs, it is not a trivial task to achieve the uniform crystal orientation in the entire device area. However, BPs with a uniform crystal plane orientation have shown superior optical characteristics compared to multidomain BPs [35]. Moreover, it was found that BPs provide a way to arbitrarily steer the wavefront of reflected light by azimuthally rotating the 3D crystals [36], enabling us to fabricate various diffractive optical elements (DOEs), such as light deflectors [36] and holograms [37]. Thus, it is necessary to accomplish the tailored 3D orientation of BP crystals from the practical perspective.

In this review, we focus on recent developments to effectively produce monodomain BPs in which a specific Miller plane is uniformly oriented parallel to a substrate. Various approaches such as the assistance of external stimuli and directing the self-assembly processes are discussed with the designs, mechanisms, and potential applications. While the approaches allow monodomain BPs in a large area, they still exhibit grain boundaries between the domains. An approach that realizes macroscopic single-crystal BPs using nanopatterned templates is also discussed. Further, in contrast to the technologies that allow only crystal plane orientation to be manipulated, we discuss a new strategy to simultaneously control two crystal axes of BP I on a substrate. The method, which is referred to as field-assisted directed self-assembly process, enables us to obtain a BP I with the [110] crystal axis oriented perpendicular to the substrate and the [001] crystal axis aligned along the predesigned surface anchoring. The mechanism with its potential applications such as volume holograms is presented. In the following, we refer to the BP I and BP II with the $(h\ k\ l)$ crystal plane oriented parallel to the substrate as BP $I_{(hkl)}$ and BP $II_{(hkl)}$, respectively.

2. Crystal Plane Orientation-Controlling Technologies

In many researches, BPs are regarded as an isotropic material with scalar electro-optic properties, thus less attention has been paid to the importance of controlling the crystal orientation. However, considering that the wavelength of Bragg reflection depends on the Miller planes (according to Equation (1)) and small platelet domains greatly degrade the performance of BP-based optical devices, it is necessary to obtain the tailored 3D orientation of BP crystals on a substrate. Belyakov et al. revealed that the optical rotatory power of a single-crystal BP is significantly enhanced compared to the polycrystalline BPs [5]. Moreover, enlarging the domain size of polymer-stabilized BPs (PSBPs) enables hysteresis-free electro-optical switching due to the reduced grain boundaries between the different crystal orientations [38]. Thus, controlling the crystal orientation and understanding the mechanisms are important from the fundamental and practical perspective. In the following, we discuss recent developments to control the orientation of BP crystals over a large area, including thermal control, the use of alignment layers, electric field treatment, and the use of nanopatterned templates. It should be noted that we refer to BPs consisting of small crystalline platelets with a uniform crystal plane orientation as monodomain BPs, which still have grain boundaries between the platelets. The single-crystal BPs indicate those consisting of a uniform crystal plane and an azimuthal orientation without the grain boundaries.

2.1. Thermal Control

Chen et al. investigated a thermal recycling process to obtain enlarged domains of BPs with a uniform crystal orientation [38]. By repeatedly performing the cooling and heating process in a temperature range where BP I appears, the heterogeneous nucleation (nucleation occurring at substrate surface) becomes dominant over the homogeneous nucleation (nucleation occurring in the bulk) during the crystal growth. As a result, the uniform BP $I_{(220)}$ was obtained with a domain size enlarged by approximately five times compared to the initial state. The uniform BP I showed significantly improved electro-optical hysteresis and higher transmittance due to the reduced light scattering, demonstrating the capability of using them as photonic applications such as displays and tunable LC lenses.

Extremely large monocrystalline BPs were attained by controlling the self-assembly process of BPs [39]. A LC mixture was cooled from the isotropic state to a temperature where BP I and BP II appear, and held for ten hours to induce the reassembly of the BP platelets, as shown in the polarized optical microscope (POM) images in Figure 2a,b. Interestingly, as a result of observing the crystal growth of BPs over time, only the domains of the BP II were reassembled and grown larger, whereas no significant change was observed in the domains of the BP I. This originates from the topological differences of the lattice structure and dislocation lines between BP I and BP II. In a unit cell of BP II, all dislocation lines are intersected at the center of the unit cell, playing as a single defect. Thus, there exists only one minimum free energy for the formation of a monocrystal. On the other hand, the BP I makes it difficult to form a monocrystal because the dislocation lines that are separated from each other have many minimum free energies. Despite the difficulty of forming a large BP I domain using a direct self-reassembly process, two approaches were proposed: using a well-aligned BP II as a precursor for monocrystal BP I and a gradient-temperature scanning (GTS) technique. A BP I with approximately mm^2-sized monodomains was obtained by rapidly cooling from enlarged BP II domains and holding the temperature for 24 h. Further, for the GTS technique, the monocrystal BP I placed on a low T-stage was exploited as a seed for nucleation (Figure 2c). When pushing the sample towards the low T-stage, BP I was grown from the seed with a uniform crystal plane orientation (Figure 2d). By adjusting the pushing direction two-dimensionally, a BP crystal having the size of 1200 × 700 μm² with a thickness of ~300 μm was obtained. The monocrystal BP I showed improved Bragg reflectivity, narrow bandwidth, and electrical tunability over a wide range of wavelengths (~100 nm), demonstrating its ability as versatile 3D photonic crystal devices.

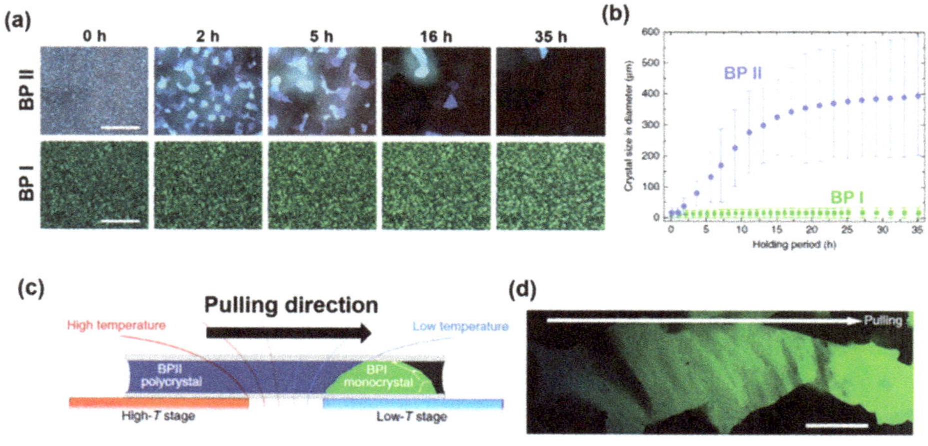

Figure 2. (**a**) Time evolutions of microscopic images of BP II (upper) and BP I (bottom) crystals. (**b**) Average domain size of BP II and BP I at various holding times. (**c**) Experimental description to obtain large single-crystal BP I through gradient-temperature scanning (GTS) method. (**d**) Polarized optical microscope (POM) image of single-crystal BP I obtained by GTS method. Reproduced with permission from [39]. Copyright 2017, Nature Publishing Group.

It is possible to obtain enlarged domains of BPs using the aforementioned methods, but only a slow rate of temperature control ensures the crystal orientation to be effectively controlled. Thus, they are all time-consuming and require complex equipment to precisely control the temperature.

2.2. Use of the Alignment Layer

One of the simplest methods to obtain monodomain BPs is to coat an appropriate alignment layer on a substrate and imprint the orientational easy axis of the LC director. Many researchers have obtained monodomain BPs particularly using a rubbing treatment and demonstrated that they possess better electro-optical and photonic characteristics than multidomain BPs. For example, Nayek et al. fabricated a monodomain BP using an antiparallel rubbed cell, which showed a reduced operating voltage by 27% and hysteresis by 63% due to the enhanced Kerr constant [40]. Similarly, a well-aligned BP II was obtained by cooling from an isotropic state in a rubbed cell, providing significantly improved reflectivity with a narrow photonic band gap [41]. A polarization-independent phase modulator was also developed using monodomain PSBPs [42]. The monodomain PSBPs showed a maximum phase shift higher than multidomain PSBPs by a factor of 1.5 under a certain applied field. A well-aligned PSBP II has shown that the hysteresis for electro-optical switching was drastically reduced compared to the multidomain PSBP I [43].

Further advanced capabilities of monodomain BPs as photonic applications were also revealed. The threshold energy for laser emission from the well-aligned BP II was drastically reduced compared to multidomain BPs, as well as the full-width of half-maximum (FWHM) of the emission became narrower by four times (Figure 3a–d) [18]. The reversibly tunable emission wavelength over a range of 40 nm was realized by adjusting the lattice constant of the well-aligned BP II through the thermal control.

A diffraction grating was demonstrated by inducing the local reorientation of the LC director through the in-plane electric field (Figure 3e). The electric field induced the LC birefringence owing to the Kerr effect, causing spatial modulation of the refractive index. Because the regular arrangement of the monodomain BPs has a long coherence length, the driving voltage was reduced by 29% with the improved diffraction efficiency by 9% (Figure 3f,g) [44].

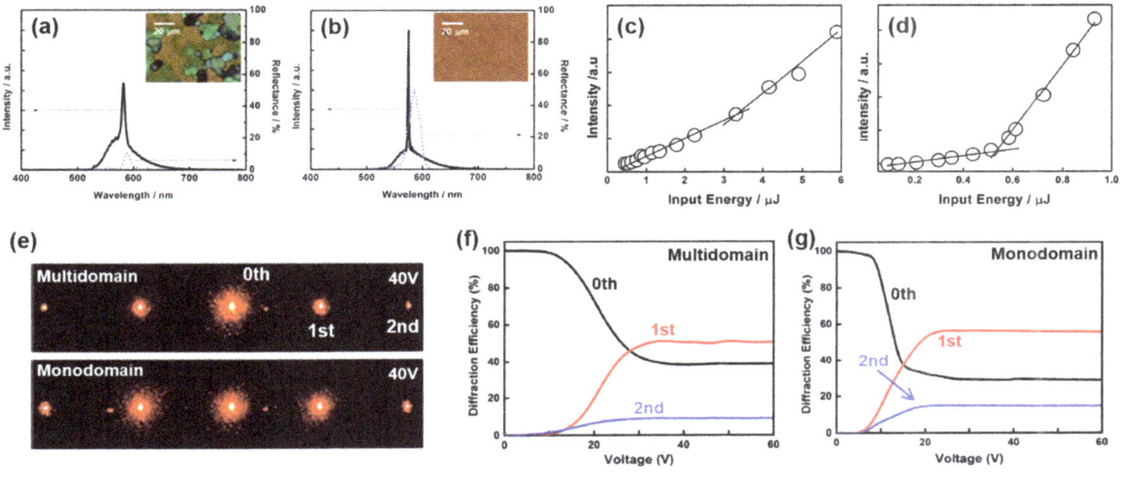

Figure 3. (**a**,**b**) Emission spectrum with POM texture (inset) of multidomain BP I (**a**) and monodomain BP II (**b**). (**c**,**d**) Output intensity with respect to the input energy of multidomain BP I (**c**) and monodomain BP II (**d**). (**e**) Far-field diffraction spots of multidomain (upper) and monodomain (lower) BP I upon applying voltage of 40 V. (**f**,**g**) Applied voltage dependent diffraction efficiencies of multidomain (**f**) and monodomain (**g**) PSBP I. (**a**–**d**) Reproduced with permission from [18]. Copyright 2015, The Royal Society of Chemistry (RSC). (**e**–**g**) Reproduced with permission from [44]. Copyright 2020, John Wiley and Sons.

The 3D orientation behavior of the BPs was investigated by observing the phase transitions of the BPs using two types of unidirectionally anchored surfaces [45]. It was revealed that the alignment of BP II obtained by cooling from the isotropic state was extremely sensitive to the surface conditions because different crystal planes, BP $II_{(110)}$ for rubbed substrate and BP $II_{(100)}$ for photo-aligned substrate, were obtained. Moreover, it was found that BPs show the strong alignment hysteresis of which the orientations of BP crystals are determined by the preceding structures. For example, BP $I_{(110)}$ becomes dominant when transited from BP $II_{(100)}$, whereas BP $I_{(211)}$ is obtained when transited from BP $II_{(110)}$. Additionally, BP I obtained from a planar cholesteric liquid crystal (ChLC) showed the (100) crystal plane oriented parallel to the substrate. This result indicates that the orientation of BPs can be easily directed by appropriately controlling the preceding LC structures.

A facile method to pattern the crystallographic orientation of BPs was developed by a photo-alignment technology [46]. A micropattern that exhibits alternate uniform (exposed region) and random (unexposed region) crystal orientations of BP II was realized by spatially illuminating a linear polarized light through a binary photo-mask as shown in Figure 4a. The BP crystals can be deformed by electric field treatment due to the external stimuli-responsivity of LCs, enabling the Bragg reflection wavelength to be tuned with a fast response time. Figure 4b shows the far-field diffraction spot from the micropatterned BP II when applying voltage of 0 V (Figure 4b(i,ii)) and 3.5 V (Figure 4b(iii,iv)). Before applying electric field, the diffraction pattern appears when illuminated by light with wavelength of 532 nm that corresponds to the Bragg reflection band of the device (Figure 4bi). On the other hand, upon application of electric voltage, the Bragg reflection band red-shifts, resulting in the diffraction pattern appearing at a longer wavelength region. Moreover, the rewritable capability of the azo-based alignment layer enables the micropattern to be switched to other patterns by illuminating light through different binary photo-masks and electric field treatment, as shown in Figure 4c. This method gives the capability to create various far-field diffraction patterns such as gratings and vortex beams through the binary reflection amplitude modulation.

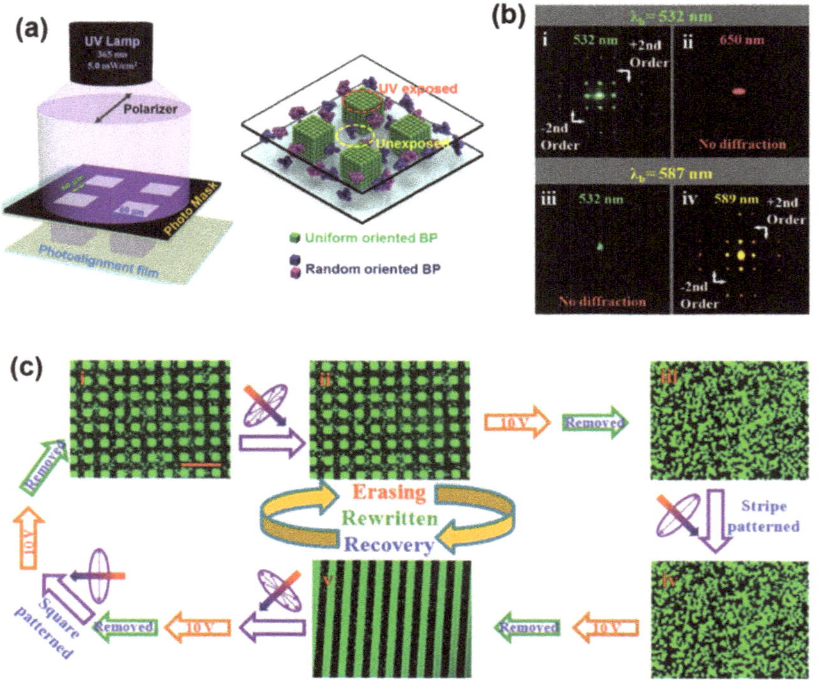

Figure 4. (**a**) Micropatterned BPs fabricated by illuminating ultraviolet (UV) light through a binary photo-mask. The region where UV light was illuminated has a uniform crystal orientation, whereas other regions have random oriented BP crystals. (**b**) External stimuli-responsive far-field diffraction patterns. (**c**) Rewritable capability of the micropatterns by illuminating non-polarized UV light and treating electric field treatment (**ii**,**iii**), and by illuminating UV light through a striped photo-mask and treating electric field treatment (**iv**,**v**). Reproduced with permission from [46]. Copyright 2017, John Wiley and Sons.

Otón et al. developed a method to predict the crystal orientation of monodomain BPs on a rubbed surface by controlling the precursor materials formulation [47]. They discovered that the crystal orientation of monodomain BPs depends on the lattice constant of BPs, and thus, the tailoring of crystal orientation can be induced by knowing the chiral twisting power of the system and adjusting the concentration of doped chiral dopants.

2.3. *Application of Electric Field*

The treatment of a vertical AC field in BPs enabled monodomain BPs to be obtained without the need of alignment layers on a substrate [48]. By applying and removing the electric field, the (110) crystal plane became dominant parallel to the substrate, resulting a monodomain-like BP textures with uniform reflection colors (Figure 5). This approach allows the crystal orientation to be maintained even when the temperature is raised to an isotropic state and then cooled to BP I. Chen et al. discussed a mechanism of the field-reorientation process of BP crystals by repeatedly applying and removing a vertical electric field [49]. When the field was applied, BPs were phase-transited to focal conic and homeotropic state. As the field was removed from the sample, the focal conic state transited to a BP with a uniform reflection color corresponding to the (110) crystal plane, whereas the homeotropic state turned into the BP $I_{(200)}$ and BP $I_{(211)}$. When the field was applied again, the portion of the focal conic state became wider than the first step, resulting in the (110) crystal plane to be dominant over a larger area as the field was removed. Thus, by repeating the electric field treatment, the growth of the different crystal planes was suppressed and only a monodomain BP $I_{(110)}$ was obtained. The monodomain BPs showed higher reflectivity with narrower bandwidth and less hysteresis compared to multidomain BPs (Figure 5b).

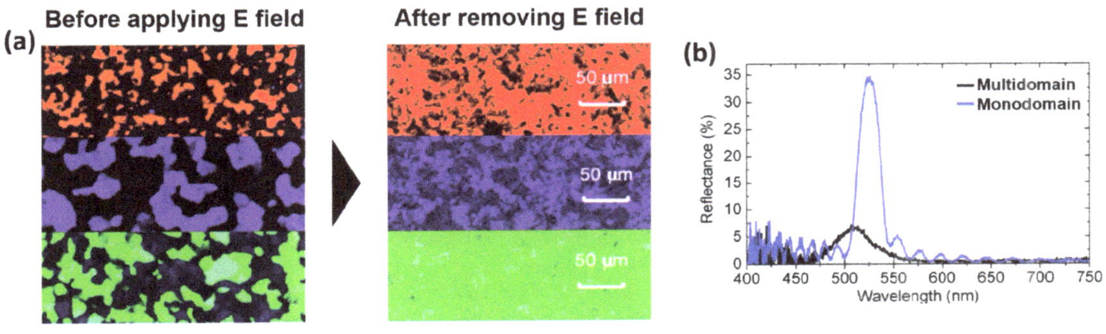

Figure 5. (**a**) POM textures of BP I before applying electric field (left) and after removing electric field (right). (**b**) Reflection spectra of multidomain and monodomain BPs. Reproduced with permission from [48], Copyright 2013, American Institute of Physics.

A field-induced bistable effect in the crystal orientation of BPs and its phase-transition mechanism were revealed using a host LC mixture with a negative dielectric anisotropy [50]. Obtaining BP I by removing the electric field from the field-induced planar ChLC state enabled a BP $I_{(200)}$ to become dominant, because the helical structure of the planar ChLC has a similar structure, with the DTC aligned perpendicular to the substrate. On the other hand, a BP $I_{(110)}$ was obtained when the field was removed from the field-induced flow state that was similar to Helfrich deformation induced by the electrohydrodynamic effect. Because the electrohydrodynamic effect increases the temperature of the sample, the field-induced flow state is phase-transited to the BP $I_{(110)}$ in a similar manner to the BP I obtained from the BP II by cooling the temperature.

A monodomain BP was obtained by initiating and growing the nuclei of BP crystals during an application of electric field [51]. When the electric field is applied at an isotropic state and the temperature of the sample is decreased to initiate the nucleation of BP crystals, heterogeneous nucleation plays a dominant role for the growth of BP crystals, while the homogeneous nucleation is suppressed. The nuclei generated through the heterogeneous process gradually grow as the temperature decreases, providing monodomain BPs in the entire region where the field is applied. This phenomenon becomes dominant upon high field intensity at a certain field frequency.

The orientation control using electric field treatment allows us to rapidly obtain monodomain BPs over a large area, but many grain boundaries still exist between the domains. Moreover, those methods can only enable a crystallographic axis of BPs to be oriented perpendicular to the substrate (crystal planes), and other axes still remain random.

2.4. Use of Nanopatterned Substrates

A new strategy to obtain single-crystal BPs was recently proposed by directing the self-assembly process of BPs using lithographically nanopatterned substrates [52,53]. The preferred orientation of the LC molecules descripted by the scalar order parameter (S) was theoretically considered for each crystallographic orientation of BPs to minimize the free energy density on a substrate. Based on simulation results, various nanopatterned templates that periodically imposed the homeotropic (H) and planar (P) alignments were designed (Figure 6a–c). For example, a striped pattern of which the H and P alignments exist alternately and the sum of the width in two regions is equal to the lattice constant of BP II was employed to obtain a single-domain BP $II_{(100)}$ over an entire patterned area without grain boundaries (Figure 6d). The crystal orientations of the BP $II_{(110)}$ and BP $II_{(111)}$ were also demonstrated using different nanopatterned templates designed in the same manner, as shown in Figure 6b,c, respectively. This method enables the single-crystal BPs to be grown up to a device thickness of approximately 20 μm [54]. The single-crystal BP II showed a single and sharp reflection peak compared to polycrystalline BPs that exhibit multiple reflection peaks due to the nonuniform lattice orientations.

The nanopatterned templates were also employed to investigate the phase transition of BPs between two different crystallographic symmetries of single-crystal BP $I_{(110)}$ and BP $II_{(100)}$ [55]. It was revealed that the transition from the single-crystal BP $II_{(100)}$ to BP $I_{(110)}$ could be regarded as martensitic-like transformation. When the BP reaches the transition temperature of BP $II_{(100)}$–BP $I_{(110)}$, the DTCs are reorganized through the local reorientation of the LC molecules, resulting in a smooth reconfiguration of the dislocation lines in a diffusionless manner on a sub-micrometer scale. This process has a little thermal hysteresis, enabling the BP crystals to be simultaneously grown in the entire patterned region. This result gives an opportunity to investigate the crystal nucleation of BPs and control the crystal growth dynamics in soft materials.

A strategy to precisely design the grain boundaries between two single-crystal BPs that have different crystal orientations, referred to as soft heteroepitaxy, was investigated by using the lithographically nanopatterned templates [56]. A C-shaped template of which the inside had a striped pattern responsible for the orientation of BP $II_{(100)}$ and the outside had a rectangular pattern responsible for BP $II_{(110)}$ was employed to observe the grain boundaries between those of crystal planes. Despite the curved geometry, the grain boundary was stabilized and elaborately sculpted on a scale similar to the lattice constant of BPs. This method was applied to fabricating a 2D diffraction grating as an external stimuli-responsive optical device. Because of the precisely sculpted grain boundaries, the signal-to-noise ratio (SNR) of the far-field diffraction patterns was significantly improved compared to the 2D diffraction gratings fabricated by controlling the crystal orientation using photo-alignment technology [46]. This work accelerates the investigation in the nucleation and growth of BP crystals between liquid−liquid interfaces, and the fabrication of intelligent photonic crystal devices.

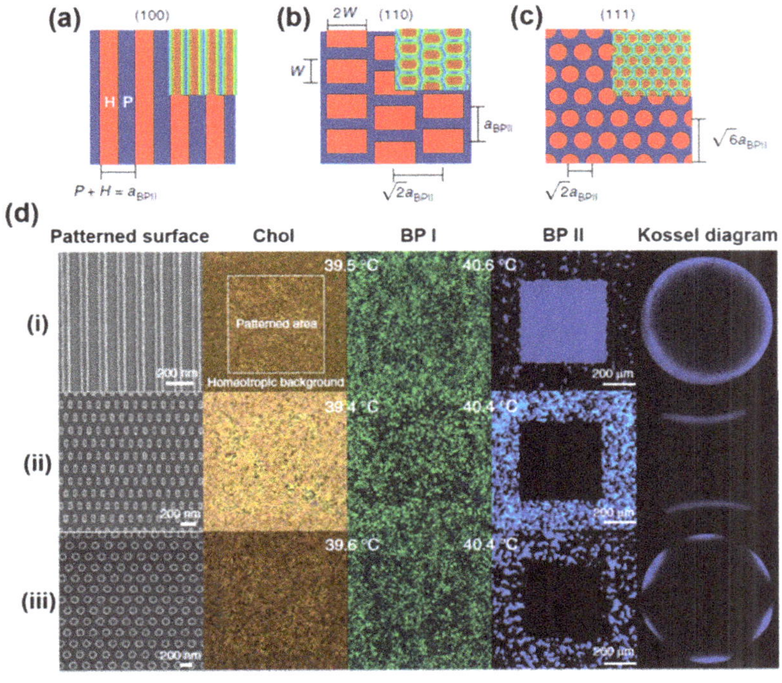

Figure 6. (**a**–**c**) Designs of lithographically nanopatterned templates for the orientation of BP $II_{(100)}$ (striped pattern), BP $II_{(110)}$ (rectangular pattern), and BP $II_{(111)}$ (circular pattern), with insets of S-map representing the preferred molecular orientation of LCs on the templates. (**d**) The scanning electron microscope image of nanopatterned templates, the POM texture of BPs, and Kossel diagrams corresponding to the BP II region, on the different templates, such as (**i**) striped pattern, (**ii**) rectangular pattern, and (**iii**) circular pattern. Reproduced with permission from [52]. Copyright 2017, Nature Publishing Group.

3. Azimuthal Orientation-Controlling Technologies of BPs

The aforementioned approaches mainly focus on obtaining monodomain-like BP textures with a uniform crystal plane orientation. However, it is required to simultaneously manipulate two independent crystal axes of BPs to uniquely define the 3D crystal orientation on a substrate. Recent studies to investigate the grain boundaries between different orientations of BP crystals [56] or realize reflective volume holograms through the Bragg–Berry (BB) effect [36] have emphasized the importance of simultaneously controlling the crystal plane and azimuthal orientation of BPs. In the following, we discuss strategies to uniquely define 3D orientation of BPs on a substrate, with its mechanisms and perspective photonic applications.

The highly ordered structures of BPs with a periodicity of a few hundred nm give rise to a unique CP-selective Bragg reflection in visible light region. With the external stimuli-responsivity of LCs, many tunable devices have been proposed by modulating either the intensity or wavelength of Bragg reflected light from BPs. On the other hand, it was recently found that a 0–2π full wavefront control of reflected light can be attained by azimuthally rotating BP crystals on a substrate, which is referred to as the BB effect. The simulations revealed that both BP $I_{(110)}$ and BP $II_{(100)}$ show the BB effect in the same manner [36]. The theoretical prediction was experimentally demonstrated by spatially rotating BP II crystals on a substrate through photo-alignment technology. The orientation of the BP $II_{(100)}$ was manipulated with a [100] axis (azimuthal orientation) to follow a predesigned surface anchoring and another [100] axis (crystal plane orientation) was aligned along the substrate normal. Whether the wavefront was indeed being controlled was confirmed through the observation of an interferometric microscope image. This result demonstrates that a full 0–2π phase-only modulation with a CP selectivity provides opportunities for advanced DOEs, such as flat deflectors, lenses, and holograms.

While the azimuthal orientation of the BP $II_{(100)}$ can be made to follow the imprinted surface anchoring on a substrate, it had still remained a challenge to attain the tailored azimuthal orientation of BP I on a substrate. Recently, a strategy to control the azimuthal orientation of BP I was proposed using a field-induced directed self-assembly process [37]. At a BP $I_{(211)}$ prepared by cooling from the isotropic state, an electric field was applied to unwind the twist structures and gradually reduced to reorient the BP crystals. A BP I with a different reflection color compared to the initial BP $I_{(211)}$ was obtained through the phase transitions of homeotropic–chiral nematic–BP X–BP I (Figure 7a). The two-fold symmetry of the Kossel diagrams revealed that the field-reoriented BP I has a (110) crystal plane parallel to the substrate (Figure 7b). Further, the Kossel patterns were rotated with a step of 45°, indicating that the azimuthal orientation ([001] crystal axis) was also controlled to follow the imprinted surface anchoring. These results showed that two crystal axes of BP I can be simultaneously controlled along the predesigned directions. The mechanism was investigated by observing the two field-induced intermediate phases, chiral nematic and BP X. The chiral nematic phase plays a key role to control the [001] axis along the predesigned direction. The field-induced chiral nematic was composed of the uniformly lying helix (ULH) with the helical axis perpendicular to the surface anchoring with LC molecules aligned along the surface anchoring on a substrate, which is a similar structure to a DTC. An additional twist of the LC molecules to create the DTC from the ULH was formed perpendicular to those of the surface anchoring and helical axis, making the DTC to be aligned along the surface anchoring. On the other hand, the field-induced BP X was responsible for the (110) crystal plane orientation of the field-induced BP I. This result was consistent with the simulation results which revealed that BP X can be obtained via continuous reorganization from BP $I_{(110)}$ [57,58].

The field-assisted directed self-assembly process was applied to a fabrication of a phase-only BP-based hologram based on the BB effect. The unique 3D chiral structure of BPs gave a capability to play back the encoded light information for only a CP light in a wide range of incident angles (Figure 7c), which were distinct from the holograms realized by ChLCs [59,60]. The holograms were expected to have significantly high light-conversion efficiency compared to binary-amplitude DOEs made by spatial control of only crystal plane orientation of BPs [12,46,56].

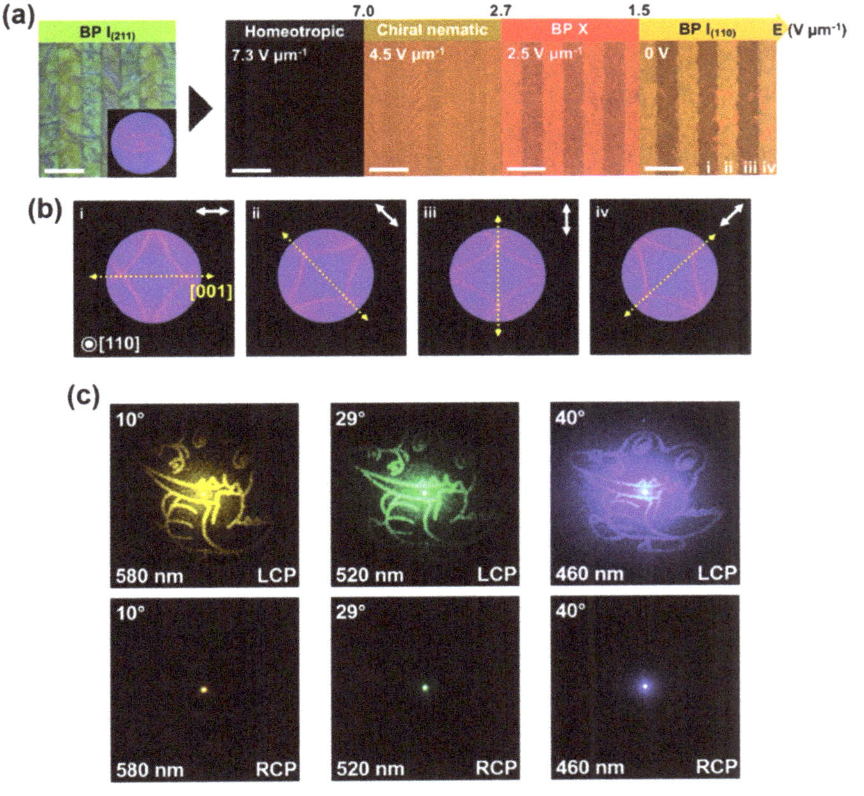

Figure 7. (**a**) The evolution of POM images of the field-induced phase transition when the field is applied to H phase and gradually reduced. (**b**) Kossel diagrams of the field-induced BP I at each striped region marked as the Roman numeral in (**a**). The double-sided white arrows and the yellow arrows show the imprinted surface anchoring in each region and the [001] crystal axis of the field-induced BP I. (**c**) Holograms reconstructed at incident angles of 10°, 29°, and 40° for left (L) and right (R) CP light illumination with wavelength of 580 nm, 520 nm, and 460 nm, respectively. Reproduced with permission from [37]. Copyright 2021, Nature Publishing Group.

4. Discussion

In this paper, we have drawn recent research progress to direct the 3D orientation of two types of cubic BPs: body-centered BP I and simple cubic BP II, along the predesigned directions on a substrate. Various strategies have emerged to manipulate either the crystal plane and azimuthal orientation of BPs, such as directing the self-assembly process and external field treatments. Controlling the 3D orientation of the BPs is one of the important steps in both science and industry fields, because it provides a degree of freedom to manipulate the light propagating through the medium, giving tailored optical characteristics in a thickness of a sub-micrometer scale. By combining the external stimuli-responsivity of BPs, various photonic devices that have not been discovered yet can be developed with high-levels and integrated optical functionalities. For example, we expect that 3D emission direction-controllable photonic crystal lasers can be achieved by introducing a slanted crystal plane orientation and controlling its periodicity [61]. Moreover, BPs have a great potential to fabricate flexible films by photo-polymerizing the LC materials [62–65], making them applicable to wearable devices such as smart-glasses. Although there still remain several questions that should be addressed such as the orientation behavior of the

LC molecules in grain boundaries, it is manifest that BPs have a great deal of potential in various scientific research field and the development of practical devices in a wide range from displays to holographic optical elements.

Author Contributions: Writing—original draft, S.C.; writing—review and editing, S.C. and M.O.; supervision, M.O. Funding acquisition, M.O. All authors have read and agreed to the published version of the manuscript.

Funding: This work was supported by the JSPS KAKENHI (20H00391, and 20H04672) and the JSPS Core-to-Core Program A., Advanced Research Networks.

Data Availability Statement: This study did not report any data.

Conflicts of Interest: The authors declare no conflict of interest.

References

1. Kikuchi, H. Liquid crystalline blue phases. *Struct. Bond.* **2007**, *128*, 99–117.
2. Higashiguchi, K.; Yasui, K.; Kikuchi, H. Direct observation of polymer-stabilized blue phase I structure with confocal laser scanning microscope. *J. Am. Chem. Soc.* **2008**, *130*, 6326–6327. [CrossRef]
3. Tanaka, S.; Yoshida, H.; Kawata, Y.; Kuwahara, R.; Nishi, R.; Ozaki, M. Double-twist cylinders in liquid crystalline cholesteric blue phases observed by transmission electron microscopy. *Sci. Rep.* **2015**, *5*, 16180. [CrossRef] [PubMed]
4. Yoshida, H.; Anucha, K.; Ogawa, Y.; Kawata, Y.; Ozaki, M.; Fukuda, J.-I.; Kikuchi, H. Bragg reflection band width and optical rotatory dispersion of cubic blue-phase liquid crystals. *Phys. Rev. E* **2016**, *94*, 042703. [CrossRef]
5. Belyakov, V.; Demikhov, E.; Dmitrienko, V.; Dolganov, V. Optical activity, transmission spectra, and structure of blue phases of liquid crystals. *J. Exp. Theor. Phys.* **1985**, *89*, 2035–2051.
6. Hur, S.T.; Lee, B.R.; Gim, M.J.; Park, K.W.; Song, M.H.; Choi, S.W. Liquid-Crystalline Blue Phase Laser with Widely Tunable Wavelength. *Adv. Mater.* **2013**, *25*, 3002–3006. [CrossRef]
7. Wang, L.; Li, Q. Stimuli-directing self-organized 3D liquid-crystalline nanostructures: From materials design to photonic applications. *Adv. Funct. Mater.* **2016**, *26*, 10–28. [CrossRef]
8. Chen, C.-W.; Li, C.-C.; Jau, H.-C.; Yu, L.-C.; Hong, C.-L.; Guo, D.-Y.; Wang, C.-T.; Lin, T.-H. Electric field-driven shifting and expansion of photonic band gaps in 3D liquid photonic crystals. *ACS Photonics* **2015**, *2*, 1524–1531. [CrossRef]
9. Du, X.-W.; Hou, D.-S.; Li, X.; Sun, D.-P.; Lan, J.-F.; Zhu, J.-L.; Ye, W.-J. Symmetric continuously tunable photonic band gaps in blue-phase liquid crystals switched by an alternating current field. *ACS Appl. Mater. Interfaces* **2019**, *11*, 22015–22020. [CrossRef] [PubMed]
10. Guo, D.-Y.; Chen, C.-W.; Li, C.-C.; Jau, H.-C.; Lin, K.-H.; Feng, T.-M.; Wang, C.-T.; Bunning, T.J.; Khoo, I.C.; Lin, T.-H. Reconfiguration of three-dimensional liquid-crystalline photonic crystals by electrostriction. *Nat. Mater.* **2020**, *19*, 94–101. [CrossRef]
11. Xu, X.; Liu, Z.; Liu, Y.; Zhang, X.; Zheng, Z.; Luo, D.; Sun, X. Electrically switchable, hyper-reflective blue phase liquid crystals films. *Adv. Opt. Mater.* **2018**, *6*, 1700891. [CrossRef]
12. Zhou, K.; Bisoyi, H.K.; Jin, J.Q.; Yuan, C.L.; Liu, Z.; Shen, D.; Lu, Y.Q.; Zheng, Z.G.; Zhang, W.; Li, Q. Light-Driven Reversible Transformation between Self-Organized Simple Cubic Lattice and Helical Superstructure Enabled by a Molecular Switch Functionalized Nanocage. *Adv. Mater.* **2018**, *30*, 1800237. [CrossRef]
13. Chen, X.; Wang, L.; Li, C.; Xiao, J.; Ding, H.; Liu, X.; Zhang, X.; He, W.; Yang, H. Light-controllable reflection wavelength of blue phase liquid crystals doped with azobenzene-dimers. *Chem. Commun.* **2013**, *49*, 10097–10099. [CrossRef]
14. Yan, J.; Wu, S.-T.; Cheng, K.-L.; Shiu, J.-W. A full-color reflective display using polymer-stabilized blue phase liquid crystal. *Appl. Phys. Lett.* **2013**, *102*, 081102. [CrossRef]
15. Yokoyama, S.; Mashiko, S.; Kikuchi, H.; Uchida, K.; Nagamura, T. Laser emission from a polymer-stabilized liquid-crystalline blue phase. *Adv. Mater.* **2006**, *18*, 48–51. [CrossRef]
16. Cao, W.; Munoz, A.; Palffy-Muhoray, P.; Taheri, B. Lasing in a three-dimensional photonic crystal of the liquid crystal blue phase II. *Nat. Mater.* **2002**, *1*, 111–113. [CrossRef] [PubMed]
17. Petriashvili, G.; Chanishvili, A.; Zurabishvili, T.; Chubinidze, K.; Ponjavidze, N.; De Santo, M.P.; Bruno, M.D.L.; Barberi, R. Temperature tunable omnidirectional lasing in liquid crystal blue phase microspheres. *OSA Contin.* **2019**, *2*, 3337–3342. [CrossRef]
18. Kim, K.; Hur, S.-T.; Kim, S.; Jo, S.-Y.; Lee, B.R.; Song, M.H.; Choi, S.-W. A well-aligned simple cubic blue phase for a liquid crystal laser. *J. Mater. Chem. C* **2015**, *3*, 5383–5388. [CrossRef]
19. Bukusoglu, E.; Wang, X.; Martinez-Gonzalez, J.A.; de Pablo, J.J.; Abbott, N.L. Stimuli-responsive cubosomes formed from blue phase liquid crystals. *Adv. Mater.* **2015**, *27*, 6892–6898. [CrossRef]
20. Martínez-González, J.A.; Zhou, Y.; Rahimi, M.; Bukusoglu, E.; Abbott, N.L.; de Pablo, J.J. Blue-phase liquid crystal droplets. *Proc. Natl. Acad. Sci. USA* **2015**, *112*, 13195–13200. [CrossRef]
21. Chen, Y.-H.; Wang, C.-T.; Yu, C.-P.; Lin, T.-H. Polarization independent Fabry-Pérot filter based on polymer-stabilized blue phase liquid crystals with fast response time. *Opt. Express* **2011**, *19*, 25441–25446. [CrossRef] [PubMed]
22. Chen, H.-Y.; Tu, H.-Y. Optical polarization states of a liquid-crystal blue phase II. *OSA Contin.* **2019**, *2*, 478–485. [CrossRef]

23. Yan, J.; Li, Y.; Wu, S.-T. High-efficiency and fast-response tunable phase grating using a blue phase liquid crystal. *Opt. Lett.* **2011**, *36*, 1404–1406. [CrossRef] [PubMed]
24. Ge, S.-J.; Ji, W.; Cui, G.-X.; Wei, B.-Y.; Hu, W.; Lu, Y.-Q. Fast switchable optical vortex generator based on blue phase liquid crystal fork grating. *Opt. Mater. Express* **2014**, *4*, 2535–2541. [CrossRef]
25. Kikuchi, H.; Yokota, M.; Hisakado, Y.; Yang, H.; Kajiyama, T. Polymer-stabilized liquid crystal blue phases. *Nat. Mater.* **2002**, *1*, 64–68. [CrossRef] [PubMed]
26. Castles, F.; Day, F.; Morris, S.; Ko, D.; Gardiner, D.; Qasim, M.; Nosheen, S.; Hands, P.; Choi, S.; Friend, R. Blue-phase templated fabrication of three-dimensional nanostructures for photonic applications. *Nat. Mater.* **2012**, *11*, 599–603. [CrossRef]
27. Yoshida, H.; Tanaka, Y.; Kawamoto, K.; Kubo, H.; Tsuda, T.; Fujii, A.; Kuwabata, S.; Kikuchi, H.; Ozaki, M. Nanoparticle-stabilized cholesteric blue phases. *Appl. Phys. Express* **2009**, *2*, 121501. [CrossRef]
28. Draude, A.P.; Kalavalapalli, T.Y.; Iliut, M.; McConnell, B.; Dierking, I. Stabilization of liquid crystal blue phases by carbon nanoparticles of varying dimensionality. *Nanoscale Adv.* **2020**, *2*, 2404–2409. [CrossRef]
29. Hur, S.-T.; Gim, M.-J.; Yoo, H.-J.; Choi, S.-W.; Takezoe, H. Investigation for correlation between elastic constant and thermal stability of liquid crystalline blue phase I. *Soft Matter* **2011**, *7*, 8800–8803. [CrossRef]
30. Yoshizawa, A.; Sato, M.; Rokunohe, J. A blue phase observed for a novel chiral compound possessing molecular biaxiality. *J. Mater. Chem.* **2005**, *15*, 3285–3290. [CrossRef]
31. Ojima, M.; Noma, T.; Asagi, H.; Fujii, A.; Ozaki, M.; Kikuchi, H. Pinning effect of mixed cellulose ester membrane on appearance of cholesteric blue phases. *Appl. Phys. Express* **2009**, *2*, 021502. [CrossRef]
32. Noma, T.; Ojima, M.; Asagi, H.; Kawahira, Y.; Fujii, A.; Ozaki, M.; Kikuchi, H. Effects of polymer network surfaces on expansion of cholesteric blue phases temperature. *e-J. Surf. Sci. Nanotechnol.* **2008**, *6*, 17–20. [CrossRef]
33. Lin, J.-D.; Ho, Y.-L.D.; Chen, L.; Lopez-Garcia, M.; Jiang, S.-A.; Taverne, M.P.; Lee, C.-R.; Rarity, J.G. Microstructure-stabilized blue phase liquid crystals. *ACS Omega* **2018**, *3*, 15435–15441. [CrossRef] [PubMed]
34. Coles, H.J.; Pivnenko, M.N. Liquid crystal 'blue phases' with a wide temperature range. *Nature* **2005**, *436*, 997–1000. [CrossRef] [PubMed]
35. Yang, J.; Zhao, W.; He, W.; Yang, Z.; Wang, D.; Cao, H. Liquid crystalline blue phase materials with three-dimensional nanostructures. *J. Mater. Chem. C* **2019**, *7*, 13352–13366. [CrossRef]
36. Yoshida, H.; Kobashi, J. Flat optics with cholesteric and blue phase liquid crystals. *Liq. Cryst.* **2016**, *43*, 1909–1919. [CrossRef]
37. Cho, S.; Takahashi, M.; Fukuda, J.-I.; Yoshida, H.; Ozaki, M. Directed self-assembly of soft 3D photonic crystals for holograms with omnidirectional circular-polarization selectivity. *Commun. Mater.* **2021**, *2*, 39. [CrossRef]
38. Chen, H.-S.; Lin, Y.-H.; Wu, C.-H.; Chen, M.; Hsu, H.-K. Hysteresis-free polymer-stabilized blue phase liquid crystals using thermal recycles. *Opt. Mater. Express* **2012**, *2*, 1149–1155. [CrossRef]
39. Chen, C.-W.; Hou, C.-T.; Li, C.-C.; Jau, H.-C.; Wang, C.-T.; Hong, C.-L.; Guo, D.-Y.; Wang, C.-Y.; Chiang, S.-P.; Bunning, T.J. Large three-dimensional photonic crystals based on monocrystalline liquid crystal blue phases. *Nat. Commun.* **2017**, *8*, 727. [CrossRef] [PubMed]
40. Nayek, P.; Jeong, H.; Park, H.R.; Kang, S.-W.; Lee, S.H.; Park, H.S.; Lee, H.J.; Kim, H.S. Tailoring monodomain in blue phase liquid crystal by surface pinning effect. *Appl. Phys. Express* **2012**, *5*, 051701. [CrossRef]
41. Kim, K.; Kim, S.; Jo, S.-Y.; Choi, S.-W. A monodomain-like liquid-crystalline simple cubic blue phase II. *J. Inf. Disp.* **2015**, *16*, 155–160. [CrossRef]
42. Oton, E.; Netter, E.; Nakano, T.; Inoue, F. Monodomain blue phase liquid crystal layers for phase modulation. *Sci. Rep.* **2017**, *7*, 44575. [CrossRef]
43. Jo, S.-Y.; Jeon, S.-W.; Kim, B.-C.; Bae, J.-H.; Araoka, F.; Choi, S.-W. Polymer stabilization of liquid-crystal blue phase II toward photonic crystals. *ACS Appl. Mater. Interfaces* **2017**, *9*, 8941–8947. [CrossRef] [PubMed]
44. Manda, R.; Pagidi, S.; Heo, Y.J.; Lim, Y.J.; Kim, M.S.; Lee, S.H. Polymer-Stabilized Monodomain Blue Phase Diffraction Grating. *Adv. Mater. Interfaces* **2020**, *7*, 1901923. [CrossRef]
45. Takahashi, M.; Ohkawa, T.; Yoshida, H.; Fukuda, J.-I.; Kikuchi, H.; Ozaki, M. Orientation of liquid crystalline blue phases on unidirectionally orienting surfaces. *J. Phys. D Appl. Phys.* **2018**, *51*, 104003. [CrossRef]
46. Zheng, Z.G.; Yuan, C.L.; Hu, W.; Bisoyi, H.K.; Tang, M.J.; Liu, Z.; Sun, P.Z.; Yang, W.Q.; Wang, X.Q.; Shen, D. Light-patterned crystallographic direction of a self-organized 3d soft photonic crystal. *Adv. Mater.* **2017**, *29*, 1703165. [CrossRef] [PubMed]
47. Otón, E.; Yoshida, H.; Morawiak, P.; Strzeżysz, O.; Kula, P.; Ozaki, M.; Piecek, W. Orientation control of ideal blue phase photonic crystals. *Sci. Rep.* **2020**, *10*, 10148. [CrossRef]
48. Chen, Y.; Wu, S.-T. Electric field-induced monodomain blue phase liquid crystals. *Appl. Phys. Lett.* **2013**, *102*, 171110. [CrossRef]
49. Chen, M.; Lin, Y.-H.; Chen, H.-S.; Chen, H.-Y. Electrically assisting crystal growth of blue phase liquid crystals. *Opt. Mater. Express* **2014**, *4*, 953–959. [CrossRef]
50. Wang, C.-T.; Liu, H.-Y.; Cheng, H.-H.; Lin, T.-H. Bistable effect in the liquid crystal blue phase. *Appl. Phys. Lett.* **2010**, *96*, 041106. [CrossRef]
51. Yan, J.; Lin, J.; Li, Q.; Li, R.-Z. Influence of long-lasting electric field on the formation of monodomain polymer stabilized blue phase liquid crystals. *J. Appl. Phys.* **2019**, *125*, 024501. [CrossRef]
52. Martínez-González, J.A.; Li, X.; Sadati, M.; Zhou, Y.; Zhang, R.; Nealey, P.F.; De Pablo, J.J. Directed self-assembly of liquid crystalline blue-phases into ideal single-crystals. *Nat. Commun.* **2017**, *8*, 15854. [CrossRef]

53. Li, X.; Martínez-González, J.A.; Park, K.; Yu, C.; Zhou, Y.; de Pablo, J.J.; Nealey, P.F. Perfection in nucleation and growth of blue-phase single crystals: Small free-energy required to self-assemble at specific lattice orientation. *ACS Appl. Mater. Interfaces* **2019**, *11*, 9487–9495. [CrossRef] [PubMed]
54. Li, X.; Martinez-Gonzalez, J.A.; de Pablo, J.J.; Nealey, P. Thickness dependence of forming single crystal by liquid-crystalline blue phase on chemically patterned surface. *SPIE* **2018**, *10555*, 1055514.
55. Li, X.; Martínez-González, J.A.; Hernández-Ortiz, J.P.; Ramírez-Hernández, A.; Zhou, Y.; Sadati, M.; Zhang, R.; Nealey, P.F.; De Pablo, J.J. Mesoscale martensitic transformation in single crystals of topological defects. *Proc. Natl. Acad. Sci. USA* **2017**, *114*, 10011–10016. [CrossRef] [PubMed]
56. Li, X.; Martínez-González, J.A.; Guzmán, O.; Ma, X.; Park, K.; Zhou, C.; Kambe, Y.; Jin, H.M.; Dolan, J.A.; Nealey, P.F. Sculpted grain boundaries in soft crystals. *Sci. Adv.* **2019**, *5*, eaax9112. [CrossRef]
57. Alexander, G.; Marenduzzo, D. Cubic blue phases in electric fields. *Europhys. Lett.* **2008**, *81*, 66004. [CrossRef]
58. Alexander, G.; Yeomans, J. Numerical results for the blue phases. *Liq. Cryst.* **2009**, *36*, 1215–1227. [CrossRef]
59. Kobashi, J.; Yoshida, H.; Ozaki, M. Polychromatic optical vortex generation from patterned cholesteric liquid crystals. *Phys. Rev. Lett.* **2016**, *116*, 253903. [CrossRef]
60. Cho, S.; Ono, M.; Yoshida, H.; Ozaki, M. Bragg-Berry flat reflectors for transparent computer-generated holograms and waveguide holography with visible color playback capability. *Sci. Rep.* **2020**, *10*, 8201. [CrossRef]
61. Cho, S.; Yoshida, H.; Ozaki, M. Emission Direction-Tunable Liquid Crystal Laser. *Adv. Opt. Mater.* **2020**, *8*, 2000375. [CrossRef]
62. Lin, T.H.; Li, Y.; Wang, C.T.; Jau, H.C.; Chen, C.W.; Li, C.C.; Bisoyi, H.K.; Bunning, T.J.; Li, Q. Red, green and blue reflections enabled in an optically tunable self-organized 3D cubic nanostructured thin film. *Adv. Mater.* **2013**, *25*, 5050–5054. [CrossRef]
63. Yang, J.; Liu, J.; Guan, B.; He, W.; Yang, Z.; Wang, J.; Ikeda, T.; Jiang, L. Fabrication and photonic applications of large-domain blue phase films. *J. Mater. Chem. C* **2019**, *7*, 9460–9466. [CrossRef]
64. Yang, J.; Zhao, W.; Yang, Z.; He, W.; Wang, J.; Ikeda, T.; Jiang, L. Printable photonic polymer coating based on a monodomain blue phase liquid crystal network. *J. Mater. Chem. C* **2019**, *7*, 13764–13769. [CrossRef]
65. Castles, F.; Morris, S.; Hung, J.; Qasim, M.M.; Wright, A.; Nosheen, S.; Choi, S.; Outram, B.; Elston, S.; Burgess, C. Stretchable liquid-crystal blue-phase gels. *Nat. Mater.* **2014**, *13*, 817–821. [CrossRef] [PubMed]

symmetry

MDPI

Article

Enhancement of Polymer Structural Ordering in Polymer-Stabilised Blue Phases for Improved Electro-Optical Properties

Hirotsugu Kikuchi [1,*], Takahiro Ashimine [2], Zehui Qin [2], Hiroki Higuchi [1], Shizuka Anan [1] and Yasushi Okumura [1,*]

1 Institute for Material Chemistry and Engineering, Kyushu University, Kasuga, Fukuoka 816-8580, Japan; hiroki19751031@gmail.com or higuchi@cm.kyushu-u.ac.jp (H.H.); anan@cm.kyushu-u.ac.jp (S.A.)

2 Interdisciplinary Graduate School of Engineering, Kyushu University, Kasuga, Fukuoka 816-8580, Japan; asshiimine@gmail.com or ashimine0@asem.kyushu-u.ac.jp (T.A.); q15521161704@163.com or qin.zehui.189@s.kyushu-u.ac.jp (Z.Q.)

* Correspondence: kikuchi@cm.kyushu-u.ac.jp (H.K.); okumura@cm.kyushu-u.ac.jp (Y.O.)

Abstract: Polymer-stabilised blue phase (PSBP) could be employed in novel fast response optical and photonic devices. It is inferred that inside PSBPs, the polymers are selectively aggregated by location in $-1/2$ disclinations, which are defects coexisting with the blue phase as a periodic lattice, thereby extending the temperature range of the blue phase. The polymer aggregate structure in PSBPs strongly affects their physical properties. In this study, we employed a non-destructive synchrotron ultra-small-angle X-ray diffraction analysis to investigate the effect of polymerisation rates on the polymer aggregate structure in PSBPs prepared with monomers of different polymerisation rates and examined the structure formation process of the polymer during polymerisation. When methacrylate monomers, which exhibit a relatively low polymerisation rate, were used to form polymers in PSBP, the resulting polymer was more selectively aggregated at disclinations in the PSBP. Furthermore, the electro-optical effect in the PSBP was successfully improved by reducing the polymer concentration in the PSBPs prepared with the optimised monomer combinations.

Keywords: liquid crystal; blue phase; polymer-stabilised blue phase; Kerr effect; electro-optics; disclination

Citation: Kikuchi, H.; Ashimine, T.; Qin, Z.; Higuchi, H.; Anan, S.; Okumura, Y. Enhancement of Polymer Structural Ordering in Polymer-Stabilised Blue Phases for Improved Electro-Optical Properties. *Symmetry* **2021**, *13*, 772. https://doi.org/10.3390/sym13050772

Academic Editors: Shoichi Ishihara and Sadahito Uto

Received: 31 March 2021
Accepted: 23 April 2021
Published: 29 April 2021

1. Introduction

Introducing chiral symmetry into nematic liquid crystals induces twisted molecular arrangements. As the chirality increases, double-twisted arrangements, which cannot continuously fill the three-dimensional space, predominate over simple twists. The cholesteric blue phase, hereafter referred to simply as the blue phase, is a liquid-crystal phase caused by the frustration between double-twisted molecular arrangements and the uniformity of the three-dimensional (3D) space, i.e., defect-free structure of molecular arrangement [1–3]. Since the blue phase is a result of the preference of the double-twisted structure over the uniformity of the 3D space, it necessarily coexists with defects, i.e., $-1/2$ disclinations. The main feature of the blue phase is that it spontaneously forms a three-dimensional periodic lattice, which has been confirmed by real space observation [4], and thereby defects also form a three-dimensional periodic lattice. The polymer-stabilised blue phase (PSBP) can be obtained by in situ polymerisation of appropriate amounts of monomers in the blue phase, and it is a liquid-crystal phase that dramatically extends the blue phase temperature range without solidification [5]. Unlike ordinary liquid crystals, PSBP is optically isotropic when no electric field is applied, and birefringence occurs when an electric field is applied. The observed electro-optical behaviour is phenomenologically similar to the Kerr effect, although the Kerr coefficient is many orders of magnitude higher than that of ordinary polar liquids due to the cooperative change in molecular orientations, which is a unique property of liquid crystals. In addition, its electro-optical response is much faster than that of ordinary nematic liquid crystals [6–9]. Therefore, PSBP could be good for applications in high-performance displays, high-speed optical modulation devices, photonic devices, etc., which are difficult to achieve with conventional nematic liquid crystals [10–14]. When an electric field is applied to PSBP and removed, the birefringence returns to zero; however, a small residual birefringence is observed sometimes [15,16]. This could be a result of insufficient stabilisation of the blue phase. Because the blue phase is easily supercooled [17], the metastable supercooled state may be misidentified as PSBP, and such a metastable state is prone to residual birefringence. Synchrotron ultra-small-angle X-ray diffraction (USAXD) measurements have revealed that polymers are selectively aggregated by location along disclinations in PSBP [18], and this aggregated structure of polymers is responsible for the stabilisation of PSBPs [19,20]. A truly stable PSBP can be obtained when the polymer correctly fills the disclinations. Controlling the polymer aggregate structure in PSBP is a good technique to improve the performance of PSBP. This study aims to develop a method for selectively aggregating polymers by location along disclinations with high order. As disclinations tend to attract particles [21], herein polymers or oligomers undergoing polymerisation, if the particles are given enough time to diffuse to reach the disclination, they would be properly trapped in the disclination. Here, we focused on the polymerisation rate of monomers and investigated its relationship with the polymer structure in PSBP by synchrotron ultra-small-angle X-ray diffraction analysis.

2. Materials and Methods

The PSBPs were prepared by irradiating a precursor solution with UV light at a temperature that indicates a blue phase. The precursor solution was composed of nematic liquid crystals, a chiral dopant, monomer, and a photoinitiator. We used an equal-weight mixture of 4-cyano-4′-pentylbiphenyl (5CB, Aldrich) and JC-1041XX (JNC Petrochemical Corp.) as the nematic liquid crystals and 2,5-bis-[4′-(hexyloxy)-phenyl-4-carbonyl]-1,4,3,6-dianhydride-D-sorbitol (ISO-$(6OBA)_2$: synthesised) as the chiral dopant. A blue phase was obtained by mixing them in appropriate ratios. In all the samples used in this study, the fraction of the chiral dopant was kept at 3.4 mole per cent of the total precursor solution. An equal-weight mixture of monofunctional nonliquid and bifunctional liquid crystalline monomers was used as the monomer to form the polymers inside the blue phase. The four types of monomers were tagged with iodine to selectively extract only the polymer structure by X-ray diffraction. The iodinated monomers were synthesised according to the procedure in the Supplementary Information of reference [18] (also see Supplementary Materials of this paper). The chemical structures of each liquid crystal and monomer are shown in Figure 1. The molar ratio of the chiral liquid crystal (liquid crystals + chiral dopant) to monomer was 91.6/8.4 in all samples. A photoinitiator of 2,2-dimethoy-2-phenyl acetophenone (Aldrich) was used to photopolymerise the monomers in the blue phase. It was added to the precursor solution at a rate of 0.1 mol% to the total solution.

Well-mixed precursor solutions of liquid crystals/chiral dopant/monomers with 0.1 mol% of photoinitiator were sandwiched between polyethylene terephthalate film substrates at a temperature where the solution is in isotropic phase (60–70 °C), and the cell gaps were fixed by 100-μm-thick film spacers to make the sample cells for X-ray measurements. The cells were irradiated with UV light with an intensity of 1.5 mW cm^{-2} and wavelength of 365 nm for 20 min to photopolymerise the monomers within the precursor solutions on a hot stage at a temperature where each solution exhibited a blue phase.

USAXD measurements were performed using a BL40B2 beamline at the synchrotron radiation facility of SPring-8 (JASRI, Hyogo, Japan). The camera length was 4297 mm, and the X-ray wavelength was 0.155 nm. Two-dimensional (2D) diffraction images at small angles were taken using an imaging plate. The diffraction intensity profiles were obtained by circular integration of the 2D diffraction images. Diffuse scatterings by the direct beam of X-ray and polyethylene terephthalate film substrates were subtracted using graphing software.

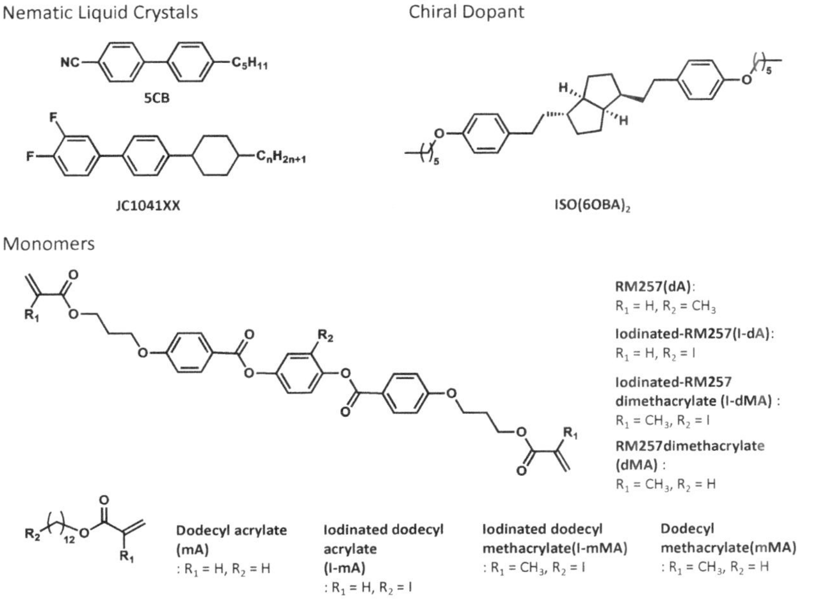

Figure 1. Chemical structures of materials for preparing PSBPs.

To determine the electro-optical properties of the PSBP, in-plane switching (IPS) cells with a 10-μm cell gap were used. The width and distance of the IPS interdigitated striped electrode were 5 and 10 μm, respectively. The cells were filled with each precursor mixture in an isotropic phase by capillary action. A system composed of a function generator (WF1974, NF Corp., Yokohama, Japan), a high voltage amplifier (2205, Trek Inc., Medina, New York, NY, USA), a differential probe (9322, HIOKI E.E. Corp., Ueda, Japan), and a data logger (GL-900-4, Graphtec Corp.) was used to apply and record an AC voltage of 1 kHz to the PSBP cells mounted on a temperature-controlled stage (10019 Peltier System, Linkam Scientific Instruments Ltd., Tadworth, UK), which was sandwiched between crossed polarisers. The data presented in this paper were measured at 23–25 °C. A white LED light source, LS-L150 (SUMITA OPTICAL GLASS, Inc., Saitama, Japan), was employed, and the changes in transmitted light due to the electric-field-induced birefringence of the PSBP were recorded as colour movies using a colour-high-speed camera (SA4, Photron Ltd., Tokyo, Japan). The transmitted light intensity was calculated using the green brightness component from the colour movie taken by the high-speed camera. The peak wavelength and half-width of spectral sensitivity curve, which is the product of the light source spectrum and the transmission spectrum of the green filter in the colour-high-speed camera, were 545 and 70 nm, respectively.

3. Results and Discussion

3.1. The Monomer Species Dependence of USAXD

Polymerisation rates depend on monomer species, such as acrylate and methacrylate. In general, the radical polymerisation rate is higher for acrylate monomers than for methacrylate monomers. This is also the case even inside PSBPs [22]. PSBPs can be prepared using a mixture of two monomers of monoacrylate with alkyl chains, such as dodecyl acrylate and diacrylate having a mesogenic core such as RM257. In this study, we used two types of monofunctional monomers (iodinated dodecyl acrylate and iodinated dodecyl methacrylate) and two types of difunctional monomers (iodinated RM-257 diacrylate and iodinated RM257 dimethacrylate). Iodine was chemically added to each monomer to increase the X-ray diffraction intensity and enhance the polymer structural contrast in

PSBP, according to Reference [18]. The chemical structures of the samples are shown in Figure 1. Five types of samples with different combinations of monomers were prepared for the USAXD measurements. The combinations are listed in Table 1.

Table 1. Combinations of monomers for preparing PSBPs.

Sample No.	Sample Symbol	Mono-Functional Monomer	Difunctional Monomer
1	mA/dA	Dodecyl acrylate	RM257
2	I-mA/I-dA	Iodinated dodecyl acrylate	Iodinated-RM257
3	I-mA/I-dMA	Iodinated dodecyl acrylate	Iodinated-RM257 dimethacrylate
4	I-mMA/I-dA	Iodinated dodecyl methacrylate	Iodinated-RM257
5	I-mMA/I-dMA	Iodinated dodecyl methacrylate	Iodinated-RM257 dimethacrylate

Figure 2 shows the USAXD 2D images trimmed in a small-angle region for PSBPs prepared using (a) mA/dA, (b) I-mA/I-dA, (c) I-mA/I-dMA, (d) I-mMA/I-dA, and (e) I-mMA/I-dMA, which correspond to samples **1**, **2**, **3**, **4** and **5**, respectively. The images were taken at room temperature around 20 °C using an imaging plate. The unexposed circular area at the centre of each image is the shadow of a beam stopper with a diameter of 2 mm. There were only very weak diffractions in sample **1**, as shown in Figure 2a. On the other hand, clear Debye rings were observed in iodinated samples (Figure 2b–e). There were clearer and stronger Debye rings for samples **4** and **5**, whereas, in samples **2** and **3**, the rings were weaker, and large central diffuse scatterings were observed. These results revealed that the polymers in samples **4** and **5** exhibited higher structural order than those in samples **2** and **3**, and a considerable amount of polymers with the nonperiodic structure were present in samples **2** and **3**.

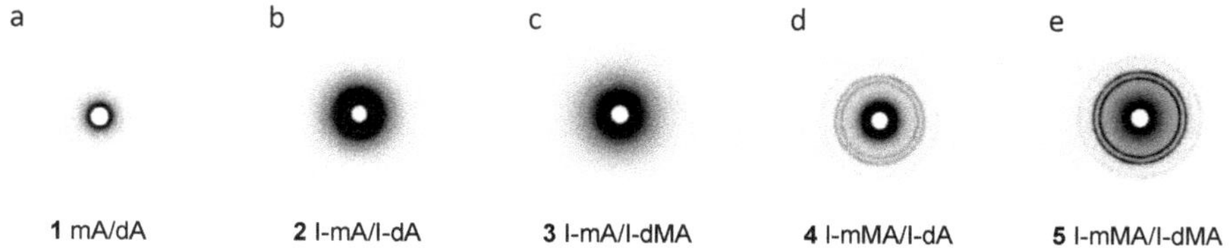

Figure 2. Dependence of monomer combinations in PSBPs prepared using (**a**) mA/dA, (**b**) I-mA/I-dA, (**c**) I-mA/I-dMA, (**d**) I-mMA/I-dA and (**e**) I-mMA/I-dMA, which correspond to samples **1**, **2**, **3**, **4** and **5**, respectively, on a USAXD 2D image.

Figure 3 shows the X-ray diffraction intensity profiles for each sample. In the body-centred cubic lattice, which is the symmetry of blue phase I, diffractions occur in the order of 110, 200, 211, 220, 310, 222, and 321 from small angles. However, in the disclination lattice within blue phase I, 110, 200, 310, and 222 disappear, according to the extinction law as calculated from its structure factor, and 211, 220, and 321 appear as diffractions. Because the observed diffraction peaks around 0.080–0.085, 0.095–0.100, and 0.125–0.130 Bragg degree are assigned to 211, 220, and 321, respectively, using the same analysis method as in the literature [18], the peaks shown in Figure 3 are attributed to the disclination lattice of blue phase I. This implies that the polymers were selectively aggregated into periodically aligned disclinations within blue phase I. The diffraction peaks of sample **5** are sharper than those of other samples, which implies that highly ordered polymers with an aggregated structure closely fitted in the disclination lattice of blue phase I were formed in sample

5. Polymers with more than a certain molecular weight are attracted to and trapped in disclination lines [21]. During polymerisation for blue phase stabilisation, an increase in the molecular weight of the polymer and a diffusion of the polymers occur simultaneously. It could take a tangible amount of time for the polymer to reach the disclinations and be trapped there. However, since cross-linking exponentially increases the molecular weight of polymers, if the polymerisation rate is too fast, diffusion will stop before reaching the disclination, resulting in a less ordered polymer structure. The polymerisation rate of acrylate monomers is much higher than that of methacrylate monomers. Therefore, in the methacrylate system, such as sample **5**, the polymerisation speed is low enough; hence, there would be enough time for the polymer to reach disclinations and form highly ordered structures. On the other hand, the polymerisation is too fast in acrylate systems, like sample **2**; thus, the molecular weight of the polymer increases rapidly and the diffusion stops before the polymer reaches disclinations. Polymers that fail to reach the disclinations form less ordered aggregates, which increases diffuse X-ray scattering.

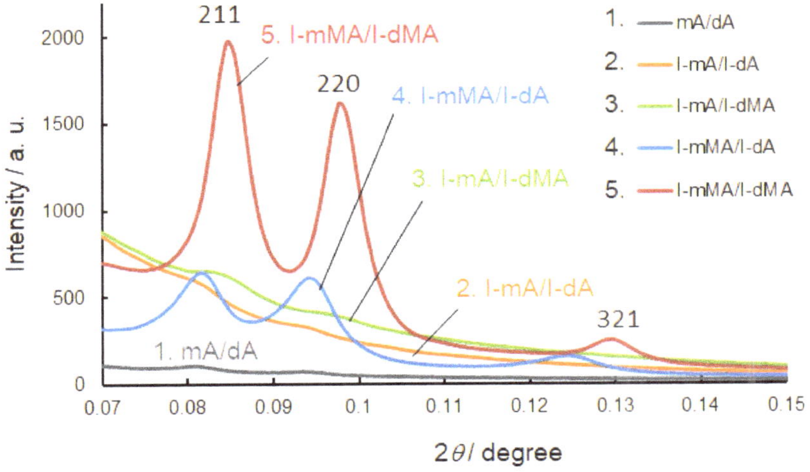

Figure 3. USAXD profiles of PSBPs prepared by using monomers with different combinations.

3.2. Temperature Dependence of USAXD

The temperature dependence of the polymer structure in PSBP was investigated for sample **5**. In a blue-phase temperature region (–40 to 46 °C), sample **5** showed a platelet texture with green and blue colours, which is a typical optical texture of the blue phase I, in a polarising optical microscope, as shown in Figure 4a. The texture turns into a dark field when heated above 46 °C due to blue–isotropic phase transition, as shown in Figure 4b. The USAXD images are shown in Figure 4c,d. Although the phases at 43 and 60 °C were completely different, their diffraction images are almost identical. This implies that the polymer structures formed along the disclination lattice in blue phase I were maintained, even after the molecular orientational order completely disappeared in the isotropic phase. This behaviour was reversible and repeatable, i.e., the same phenomenon was observed no matter how many times the phase transition was repeated by increasing and decreasing the temperature between 43 and 60 °C.

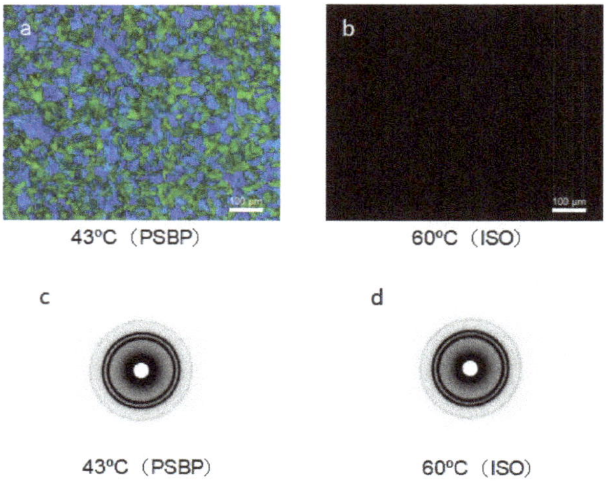

Figure 4. Polarizing optical microscope images of PSBP (sample **5**) at 43 °C (**a**) and 60 °C (**b**), and USAXD images of PSBP (sample 5) at 43 °C (**c**) and 60 °C (**d**).

On the other hand, cooling the sample changed the diffraction intensity. Figure 5 shows the USAXD profiles of sample **5** at 25.5, 0, −10, −20, −30, and −40 °C. As the temperature decreased, the diffraction peaks became smaller. In particular, the peak intensities decreased significantly between −20 and −30 °C. This suggests that the order of the polymer structure was reduced by cooling. However, when the temperature was increased back to 25 °C, the profile almost returned to its original shape. Therefore, the polymer cannot be permanently damaged by cooling, and the order of the polymer is reversibly restored by reheating to room temperature.

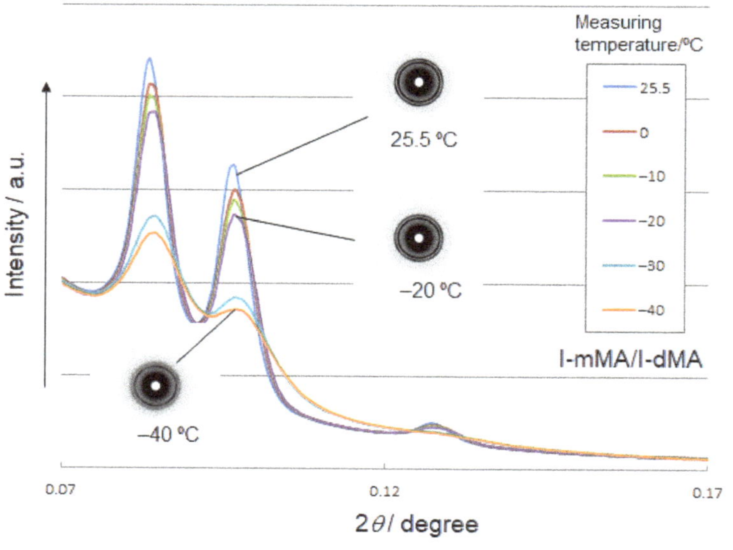

Figure 5. USAXD profiles of sample **5** at 25.5, 0, −10, −20, −30, and −40 °C.

3.3. UV Exposure-Time-Dependence of USAXD

The polymerisation of acrylate and methacrylate monomers proceeds with UV-light exposure time. Figure 6 shows the UV-light exposure-time-dependence of the USAXD profiles for sample **5** measured at 32 °C. Almost no diffraction was observed in the initial state (zero exposure time). This means that the monomers were dispersed randomly and uniformly in the blue phase before polymerisation. As the UV exposure started, the intensity of diffraction peaks gradually increased with an increase in the exposure time.

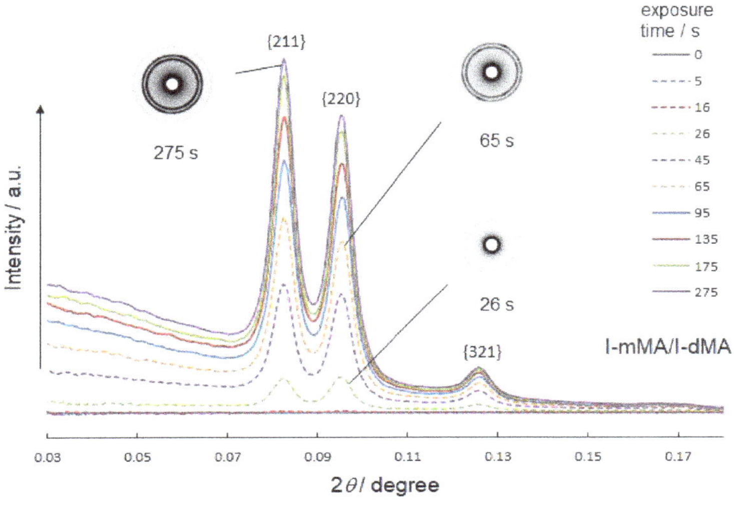

Figure 6. UV light exposure time dependence of USAXD profiles for sample **5**.

The changes in diffraction peak areas and peak half widths for 211, 220, and 321 with time during the UV exposure are shown in Figure 7a,b, respectively. Each peak area monotonically increased with exposure time and gradually saturated, whereas the half-width of each peak was almost unchanged. Therefore, we infer that as polymerisation begins, the polymers start to be trapped immediately in the disclinations, forming a highly ordered structure, and as polymerisation progresses, the polymers in the disclinations become highly aggregated, thereby increasing the contrast in the polymer concentration. Figure 8 shows the schematic illustration of a possible monomer distribution state before polymerisation in the initial and aggregated states of the polymers after polymerisation.

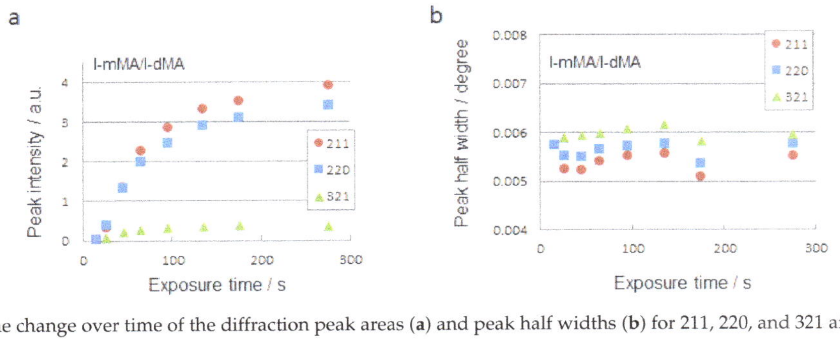

Figure 7. The change over time of the diffraction peak areas (**a**) and peak half widths (**b**) for 211, 220, and 321 after the start of UV exposure.

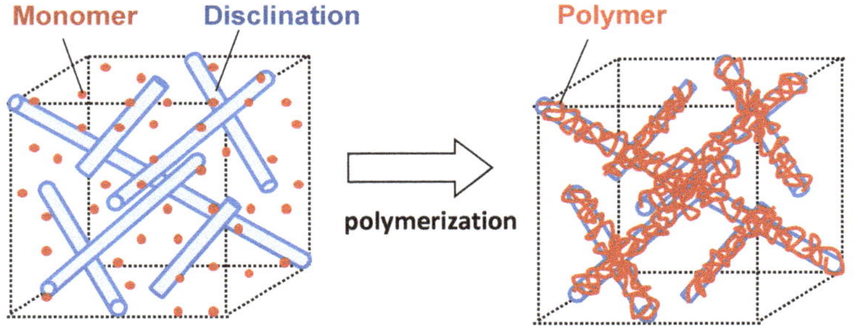

Figure 8. Schematic illustration of a possible distribution state of a monomer before polymerisation in the initial state (**left**) and an aggregation state of a polymer after polymerisation (**right**).

3.4. Mechanism of Polymer-Stabilisation of Blue Phase

In general, the temperature range of a phase is determined by the relative thermodynamic stability of the phase of interest with reference to other phases at the high- and low-temperature sides. In the case of PSBP, since the temperature range on the low-temperature side is extended, it is considered that the thermodynamic balance with the chiral nematic phase, which is on the low-temperature side, is significantly altered by the presence of a polymer. The blue phase and the chiral nematic phase conflict between double- and simple-twist structures. The blue phase is stable when the double-twist structure, which is a locally stable structure, is dominant at the expense of defect (disclination) generation, and the chiral nematic phase is stable when the simple-twist structure, which is a defect-free state, is dominant at the expense of the locally stable structure. As temperature increases and approaches the isotropic phase, defects exist more easily due to the decrease in the orientational order. Thus, a blue phase generally appears just below the isotropic phase. The experimental results obtained in this study show that when polymers are generated and grown in the blue phase, they aggregate in the disclination, a defect in the blue phase. This implies that the defects are immobilised by filling them with polymers, and the defect-free state, which is the basis of stabilisation in the chiral nematic phase, is lost. Therefore, the equilibrium state between the two phases is greatly disrupted, and the relative stability of the blue phase at lower temperatures would increase. In other words, the chiral nematic phase is replaced by the blue phase because the disadvantage of defects has become a common expense not only in the blue phase but also in the chiral nematic phase due to the immobilisation of defects. It could be inferred that breaking the spatial uniformity destabilises the chiral nematic phase. Because the blue–isotropic phase transition temperature, which is the higher temperature phase, is almost the same with and without the polymer, the polymer does not affect the ordered structure of the blue phase. This fact supports our hypothesis. In other words, the extension of the temperature range of the blue phase is not attributed to the stabilisation of the blue phase but the destabilisation of the chiral nematic phase. This can be expressed in terms of a free energy-based phase transition phenomenon, as shown in Figure 9. That is, the blue phase originally exists in a small temperature range between the chiral nematic and isotropic phases. However, when the chiral nematic phase is destabilised (i.e., the free energy of the chiral nematic phase increases), the transition temperature between the chiral nematic and blue phases shifts to a lower temperature, whereas the blue–isotropic phase transition temperature remains unchanged; thus, the temperature range of the blue phase is extended to the lower temperature side. This agrees well with experimental observations [5].

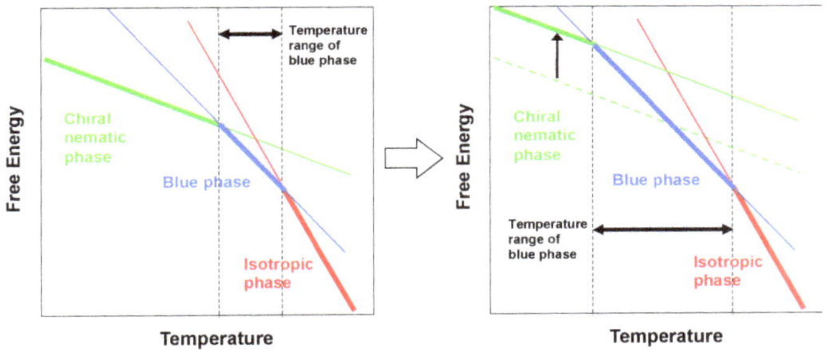

Figure 9. A possible explanation of expending blue phase temperature range in PSBP based on free energy change.

3.5. *Improvement of Electro-Optical Property of PSBP*

The blue phases I and II are optically isotropic due to their cubic structure, and when an external electric field is applied, they induce birefringence proportional to the square of the electric field; this is called the electro-optical Kerr effect. Phenomena that can be induced in the blue phases by applying an electric field include (1) phase transition to the chiral nematic phase, (2) lattice rotation, (3) lattice deformation, (4) reorientation of local directors in the lattice, and (5) changes in molecular orientation; (1–3) generally result in responses slower than milliseconds. In PSBPs, such slow responses are suppressed because the lattice is immobilised by the polymer; only fast responses, (4) and (5), occur. Since the change in birefringence resulting from (5) is small, the Kerr effect in PSBPs is mainly attributed to (4). As revealed by the USAXD experiments, inside PSBPs, there are polymers formed in a well-ordered structure along the disclinations and those aggregated in a disordered state. The latter inhibits the electro-optic effect of PSBPs and reduces the induced birefringence. Eliminating the disordered polymer aggregates in PSBPs would reduce the drive voltage, which is a major issue in PSBPs. Simply reducing the polymer fraction in PSBPs would lower the drive voltage, but there is a trade-off below which further reduction would reduce the stability of PSBPs. The decrease in the stability of PSBPs appears as residual birefringence after removing the electric field. It is essential to properly present polymers in the disclinations to prevent a decrease in the stability of PSBPs. As revealed in this study, sample **5** contains the highest ordered polymer structure, which is considered to be properly distributed along the disclinations in the PSBP. Therefore, we investigated the possibility of reducing the driving voltage while maintaining the stability of a PSBP by reducing the polymer fraction in sample **5**. Figure 10 shows the polymer-fraction-dependence of applied voltage vs. transmittance curves for PSBP prepared with dodecyl methacrylate and RM257 dimethacrylate (noniodinated). Since the transmittance was obtained by extracting and analysing the data at a wavelength of 545 nm, the transmitted light repeatedly increased and decreased due to the interference with the induced birefringence. Here, the voltage at the first maximum transmittance is defined as the driving voltage. By reducing the polymer fraction by about 13% (from 7.7 to 5.7 wt %), the driving voltage was reduced by about 29%. Almost no residual birefringence was observed in all the samples. This shows that the conventional trade-off can be overcome and the driving voltage can be effectively reduced in PSBP prepared with dodecyl methacrylate and RM257 dimethacrylate (noniodinated) where the polymer should be more properly formed in disclination. Although the measured driving voltage itself is large for a distance of 10 μm between electrodes because we used a general-purpose material as the liquid crystal, the methodology employed in this study can be applied to other liquid crystals. The driving voltage would be sufficiently reduced if a liquid-crystal material with a high Kerr coefficient developed for PSBP is used.

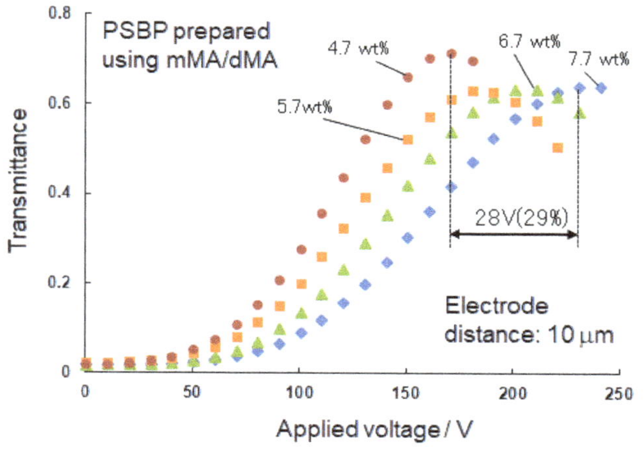

Figure 10. Polymer fraction dependence of the applied voltage vs. transmittance curves for PSBP prepared using dodecyl methacrylate and RM257 dimethacrylate (noniodinated). An electric voltage of 1 kHz was applied between electrodes 10 μm apart. The transmittance was measured at a wavelength of 545 nm.

4. Conclusions

The aggregation structures of polymers inside PSBPs were evaluated by synchrotron USAXD analyses. The PSBPs were prepared with various combinations of acrylate and methacrylate monomers with high and low polymerisation rates, respectively. Using the methacrylic monomers with low polymerisation rates resulted in the formation of highly ordered polymer aggregates that closely fit the defect lattice in the PSBP. The structure of the polymer inside the PSBP remained almost unchanged even after repeated heating of the PSBP to the isotropic phase and cooling to the original temperature. On the other hand, when the PSBP was cooled to −20 °C and below, the order of the polymer aggregation structure decreased, but when reheated to the original temperature, the polymer aggregate structure also returned to its original state. The driving voltage of the Kerr effect was effectively reduced by reducing the polymer concentration in the PSBPs prepared with methacrylate monomers, which exhibited the highest ordered polymer structure. There was almost no residual birefringence in the PSBP; hence, we conclude that the conventional trade-off between the driving voltage and residual birefringence was successfully overcome.

Supplementary Materials: Available online at https://www.mdpi.com/article/10.3390/sym13050772/s1.

Author Contributions: Conceptualisation, H.K.; methodology, H.K. and Y.O.; chemical synthetic works and chemical analyses, H.H., T.A. and S.A.; measurements and data analyses, H.H., T.A., Z.Q. and Y.O.; writing—original draft preparation, H.K.; writing—review and editing, S.A. and Y.O.; supervision, H.K. All authors have read and agreed to the published version of the manuscript.

Funding: This work was supported by the Grant-in-Aid for Scientific Research (A) JSPS KAKENHI Grant Number 18H03920 from the Japan Society for the Promotion of Science (JSPS), Dynamic Alliance for Open Innovation Bridging Human, Environment and Materials from the Ministry of Education, Culture, Sports, Science and Technology, Japan (MEXT), the Cooperative Research Program of "Network Joint Research Center for Materials and Devices" and CREST, JST (JPMJCR1424).

Acknowledgments: The synchrotron radiation experiments were performed at the BL40B2 of SPring-8 with the approval of the Japan Synchrotron Radiation Research Institute (JASRI) (Proposal No: 2014B1408). The authors thank Hiroyasu Masunaga and Noboru Ohta for their technical help with

X-ray measurements at SPring8. The authors thank JNC Co. Ltd. for providing the nematic mixture JC-1041XX.

Conflicts of Interest: The authors declare no conflict of interest.

References

1. Wright, D.C.; Mermin, N.D. Crystalline Liquids: The Blue Phases. *Rev. Mod. Phys.* **1989**, *61*, 385–432. [CrossRef]
2. Kikuchi, H. Liquid crystalline blue phases. In *Liquid Crystalline Functional Assemblies and Their Supramolecular Structures*; Kato, T., Ed.; Structure and Bonding; Springer: Berlin, Germany, 2008; Volume 128, pp. 99–117.
3. Meiboom, S.; Sethna, J.P.; Anderson, P.W.; Brinkman, W.F. Theory of the Blue Phase of Cholesteric Liquid-Crystals. *Phys. Rev. Lett.* **1981**, *46*, 1216–1219. [CrossRef]
4. Higashiguchi, K.; Yasui, K.; Kikuchi, H. Direct observation of polymer-stabilized blue phase I structure with confocal laser scanning microscope. *J. Am. Chem. Soc.* **2008**, *130*, 6326. [CrossRef] [PubMed]
5. Kikuchi, H.; Yokota, M.; Hisakado, Y.; Yang, H.; Kajiyama, T. Polymer-stabilized liquid crystal blue phases. *Nat. Mater.* **2002**, *1*, 64–68. [CrossRef] [PubMed]
6. Kikuchi, H.; Hisakado, Y.; Uchida, K.; Nagamura, T.; Kajiyama, T. Fast electro-optical effect in polymer-stabilized blue phases. In *Liquid Crystals VIII*; Khoo, I.C., Ed.; Proceedings of SPIE; SPIE: Bellingham, WA, USA, 2004; Volume 5518, pp. 182–189.
7. Hisakado, Y.; Kikuchi, H.; Nagamura, T.; Kajiyama, T. Large electro-optic Kerr effect in polymer-stabilized liquid-crystalline blue phases. *Adv. Mater.* **2005**, *17*, 96. [CrossRef]
8. Haseba, Y.; Kikuchi, H.; Nagamura, T.; Kajiyama, T. Large electro-optic Kerr effect in nanostructured chiral liquid-crystal composites over a wide temperature range. *Adv. Mater.* **2005**, *17*, 2311. [CrossRef]
9. Choi, H.; Higuchi, H.; Kikuchi, H. Fast electro-optic switching in liquid crystal blue phase II. *Appl. Phys. Lett.* **2011**, *98*, 3. [CrossRef]
10. Chen, Y.; Wu, S.T. Recent Advances on Polymer-Stabilized Blue Phase Liquid Crystal Materials and Devices. *J. Appl. Polym. Sci.* **2014**, *131*, 40556. [CrossRef]
11. Shibayama, S.; Higuchi, H.; Okumura, Y.; Kikuchi, H. Dendron-Stabilized Liquid Crystalline Blue Phases with an Enlarged Controllable Range of the Photonic Band for Tunable Photonic Devices. *Adv. Funct. Mater.* **2013**, *23*, 2387–2396. [CrossRef]
12. Yokoyama, S.; Mashiko, S.; Kikuchi, H.; Uchida, K.; Nagamura, T. Laser emission from a polymer-stabilized liquid-crystalline blue phase. *Adv. Mater.* **2006**, *18*, 48–51. [CrossRef]
13. Cao, W.Y.; Munoz, A.; Palffy-Muhoray, P.; Taheri, B. Lasing in a three-dimensional photonic crystal of the liquid crystal blue phase II. *Nat. Mater.* **2002**, *1*, 111–113. [CrossRef]
14. Yoshizawa, A. Material design for blue phase liquid crystals and their electro-optical effects. *RSC Adv.* **2013**, *3*, 25475–25497. [CrossRef]
15. Mizunuma, T.; Oo, T.N.; Nagano, Y.; Ma, H.Y.; Haseba, Y.; Higuchi, H.; Okumura, Y.; Kikuchi, H. Electro-optical properties of polymer-stabilized blue phase with different monomer combination and concentration. *Opt. Mater. Express* **2011**, *1*, 1561–1568. [CrossRef]
16. Oo, T.N.; Mizunuma, T.; Nagano, Y.; Ma, H.Y.; Ogawa, Y.; Haseba, Y.; Higuchi, H.; Okumura, Y.; Kikuchi, H. Effects of monomer/liquid crystal compositions on electro-optical properties of polymer-stabilized blue phase liquid crystal. *Opt. Mater. Express* **2011**, *1*, 1502–1510. [CrossRef]
17. Choi, H.; Higuchi, H.; Ogawa, Y.; Kikuchi, H. Polymer-stabilized supercooled blue phase. *Appl. Phys. Lett.* **2012**, *101*, 5. [CrossRef]
18. Kikuchi, H.; Izena, S.; Higuchi, H.; Okumura, Y.; Higashiguchi, K. A giant polymer lattice in a polymer-stabilized blue phase liquid crystal. *Soft Matter* **2015**, *11*, 4572–4575. [CrossRef] [PubMed]
19. Fukuda, J. Stabilization of a blue phase by a guest component: An approach based on a Landau-de Gennes theory. *Phys. Rev. E* **2010**, *82*, 5. [CrossRef] [PubMed]
20. Fukuda, J. Stability of cholesteric blue phases in the presence of a guest component. *Phys. Rev. E* **2012**, *86*, 6. [CrossRef] [PubMed]
21. Higashiguchi, K.; Yasui, K.; Ozawa, M.; Odoi, K.; Kikuchi, H. Spatial distribution control of polymer nanoparticles by liquid crystal disclinations. *Polym. J.* **2012**, *44*, 632–638. [CrossRef]
22. Iwata, T.; Suzuki, K.; Amaya, N.; Higuchi, H.; Masunaga, H.; Sasaki, S.; Kikuchi, H. Control of Cross-Linking Polymerization Kinetics and Polymer Aggregated Structure in Polymer-Stabilized Liquid Crystalline Blue Phases. *Macromolecules* **2009**, *42*, 2002–2008. [CrossRef]

symmetry MDPI

Article

Electro-Optical Characteristics of Quasi-Homogeneous Cell in Twisted Nematic Mode

Rumiko Yamaguchi * and **Yoshiki Sakamoto**

Electrical and Electronic Engineering Course, Graduate School of Engineering Science, Akita University, Akita 010-8502, Japan
* Correspondence: yrumiko@gipc.akita-u.ac.jp

Abstract: A liquid crystal (LC) director distribution was numerically analyzed in 90-degree twisted nematic (TN) LC cells with a symmetric and an asymmetric azimuthal anchoring strength of the alignment substrate and the influence of anchoring strength on the electro-optical property of the TN cell was evaluated. The twist angle decreased with decreasing azimuthal anchoring strength and the LC orientation changed to a homogeneous orientation with the twist angle of 0 degrees in the LC cell with asymmetric azimuthal anchoring strength, specifically with the strong anchoring substrate and the weak anchoring substrate below a critical strength. The asymmetric anchoring LC cell was fabricated by using a poly (vinyl cinnamate) alignment substrate as the weak anchoring surface and a polyimide alignment substrate as the strong anchoring surface. The LC cell performed the dark–bright–dark switching of the transmittance in the crossed polarizers, since the homogeneous LC orientation changed to the TN orientation again with increasing the applied voltage. Therefore, it was experimentally confirmed that LC molecules rotated at 90 degrees in the plane on the alignment surface by the electric field perpendicular to the weak anchoring substrate.

Keywords: liquid crystal; polar anchoring; azimuthal anchoring; twisted nematic; homogeneous orientation; voltage–transmittance curve

Citation: Yamaguchi, R.; Sakamoto, Y. Electro-Optical Characteristics of Quasi-Homogeneous Cell in Twisted Nematic Mode. *Symmetry* **2023**, *15*, 597. https://doi.org/10.3390/sym15030597

Academic Editors: Shoichi Ishihara and Sadahito Uto

Received: 28 December 2022
Revised: 17 February 2023
Accepted: 24 February 2023
Published: 26 February 2023

1. Introduction

Twisted nematic liquid crystal (TN LC) configurations [1] are still the most widely used form of liquid crystal display mode. The LC director distribution and electro-optical characteristics of the TN cell are generally determined by the following three features: elastic forces in the LC which are described by Oseen-Frank free-energy density [2], an electric field generated by applying a voltage to electrodes, and anchoring strength of the LC molecules on substrate boundaries [2,3]. The anchoring strength is one of the important design parameters for practical applications. Commercially available TN displays have usually been manufactured using strong azimuthal and polar anchoring surfaces at both sides of the substrate, for example, rubbed polyimide (PI) coated substrates, since the strong anchoring enables fast relaxation from field-driven states. On the other hand, an application of the weak polar anchoring interface has been proposed to decrease the driving voltage of the LC cell. The numerical analysis of the electro-optical property has been reported in the TN display mode [4–7]. Conversely, the increase in the threshold voltage has been reported if the azimuthal anchoring strength reduces by more than one order of magnitude from 2.0×10^{-3} to 5.0×10^{-5} N/m [8]. These studies have usually been carried out the same anchoring strength on both sides of the substrate.

An in-plane switching (IPS) mode [9] is another LC display mode currently used in producing televisions, monitors of personal computers, and various instruments. The IPS mode exhibits a wide viewing angle and a small color shift because LC molecules are initially homogeneously aligned and rotate within a plane parallel to the substrates when an in-plane field is applied. Weak azimuthal anchoring interfaces in the IPS mode have been reported to enlarge the allowable range of marginal variation in the cell gap [10]. Moreover, a threshold behavior has been analyzed in the IPS mode LC cell with asymmetric azimuthal anchoring strength [11]. When the weak azimuthal anchoring alignment film is coated on the interdigital electrode substrate and the counter substrate has strong anchoring strength, the LC molecules glide on the weak azimuthal anchoring surface parallel to the in-plane field direction, which results in the decrease in the driving voltage and the increase in the maximum transmission in the bright state [12–14]. The asymmetrical azimuthal anchoring configuration has also been applied to a fringe-field switching (FFS) mode [15]. Higher transmittance was obtained by lower driving voltage compared to the cell with the symmetrical anchoring configuration [16].

We numerically investigated the LC director distribution and electro-optical characteristics in a hybrid-aligned nematic cell with asymmetric anchoring strength [17]. When the planar alignment substrate with the strong polar anchoring and the homeotropic alignment with the weak polar anchoring substrate were combined, the hybrid orientation turned to the homogeneous orientation, which was called a quasi-homogeneous (Q-Homo) cell. A conventional homogeneous LC cell has a threshold voltage of about 1–2 V. However, the Q-Homo cell had no threshold voltage, and neither did the hybrid-aligned nematic cell. Moreover, a large retardation variation was also obtained compared to that in both the hybrid-aligned and homogeneous cells with symmetric strong anchoring. In a guest–host mode, a higher contrast ratio by lower driving voltage was compatible.

The 90-degree TN orientation also changed to the homogeneous orientation between strong anchoring and weak azimuthal anchoring substrates. We called the cell a quasi-homogeneous TN (Q-Homo-TN) cell. The electro-optical characteristics were theoretically estimated in the Q-Homo-TN cell and a unique voltage–transmission curve of 0–100–0% was obtained under the crossed polarizers. The infinite strength of the polar anchoring and the pretilt angle of 0 degrees were set at both sides of the substrate in the above calculations to simplify the analytical model.

In the present paper, finite values of polar and azimuthal anchoring strengths were utilized in the numerical analysis of the LC director distribution in the 90-degree TN and the Q-Homo-TN cells. The finite polar anchoring strength and the non-zero pretilt angle were also considered. Experimentally, a photo-crosslinked poly (vinyl cinnamate) (PVCi) alignment film was used as a variable azimuthal anchoring strength surface. The 90-degree TN and Q-Homo-TN cells were fabricated by adjusting the anchoring strength of the PVCi film surface. Electro-optical characteristics were investigated and voltage–transmission curves were obtained, as well as the theoretical curves.

2. Principle

2.1. Free Energy of TN LC Cell

The LC director distribution was calculated, which is based on the continuum theory. Figure 1 shows the definition of the LC director $\boldsymbol{n}$, with a tilt angle θ, and a twist angle ϕ, in the TN cell. $W_{\mathrm{p\_0}}$ and $W_{\mathrm{p\_d}}$ are the polar anchoring strengths and $W_{\mathrm{a\_0}}$ and $W_{\mathrm{a\_d}}$ are azimuthal anchoring strengths at the lower and upper side of the substrates, respectively. The total free energy per unit area, F, in the TN LC cell is given by

$$
\begin{aligned}
F &= F_{\mathrm{bulk}} + F_{\mathrm{electric}} + F_{\mathrm{surface}} \\
F_{\mathrm{bulk}} + F_{\mathrm{electric}} &= \int_0^d \tfrac{1}{2}\{(K_{11}\cos^2\theta(z) + K_{33}\sin^2\theta(z))(\tfrac{d\theta(z)}{dz})^2 + (K_{22}\cos^2\theta(z) + K_{33}\sin^2\theta(z))\cos^2\theta(z)(\tfrac{d\phi(z)}{dz})^2 \\
&\quad -\varepsilon_0(\varepsilon_\perp + \Delta\varepsilon\sin^2\theta(z))(\tfrac{dV(z)}{dz})^2\} \\
F_{\mathrm{surface}} &= \tfrac{1}{2}W_{\mathrm{p_0}}\sin^2(\theta_0 - \theta(0)) + \tfrac{1}{2}W_{\mathrm{p_d}}\sin^2(\theta_{\mathrm{d}} - \theta(d)) + \tfrac{1}{2}W_{\mathrm{a_0}}\sin^2(\phi_0 - \phi(0))\tfrac{1}{2}W_{\mathrm{a_d}}\sin^2(\phi_{\mathrm{d}} - \phi(d)),
\end{aligned}
\tag{1}
$$

where K_{11}, K_{22}, and K_{33} are the splay, twist, and bend elastic constants; $\varepsilon_\perp$ is the dielectric constant of the short axis; $\Delta\varepsilon$ is the dielectric constant anisotropy; d is the thickness of the LC layer; $V(z)$ is the voltage potential. θ_0 and ϕ_0 are polar and azimuthal easy axis angles of the lower side of the substrate, respectively. θ_{d} and ϕ_{d} are angles of easy axes at the upper side of the substrate. In this study, ϕ_0 and ϕ_d are set to 0 and 90 degrees; θ_0 and θ_d are the pretilt angles generally caused by the rubbing treatment on the polymer surface.

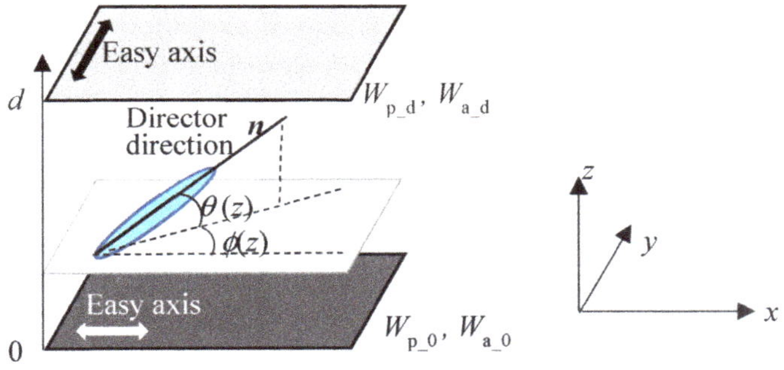

Figure 1. Definition of the LC director, ***n***, and surface alignment conditions of the substrate in the TN cell.

2.2. Critical Anchoring

The liquid crystal director distribution was obtained by minimizing the total free energy of the cell. The finite difference method was used to estimate $\theta(d)$ and $\phi(z)$ in the LC cell. Table 1 shows the physical parameters of 5CB (4-n-pentyl-4′-cyanobiphenyl) which was used in the calculation and the cell fabrication. The cell thickness was 8 μm. Figure 2a,b show the $\phi(0)$ and $\phi(d)$ as a function of azimuthal anchoring strength. The polar anchoring strengths and the tilt angles of both sides of the substrate were 1.0×10^{-3} N/m and 0 degrees, respectively. When the azimuthal anchoring strengths are the same for the substrates ($W_{a\_0} = W_{a\_d}$), $\phi(0)$ increases from 0 degrees, and $\phi(d)$ decreases from 90 degrees with decreasing the azimuthal anchoring strength, as shown in Figure 2a; $\phi(0)$ and $\phi(d)$ reached close to 45 degrees at W_{a_d} of 1.0×10^{-10} N/m and the twist angle was, however, about 0.01 degrees.

Table 1. Physical parameters of the LC.

K_{11}	K_{22}	K_{33} [pN]	$\varepsilon_{//}$	$\varepsilon_{\perp}$	n_o	n_e (550 nm)
6.3	4	8.4	17.9	6.9	1.540	1.724

On the other hand, in the TN LC cell with the asymmetrical anchoring condition with small W_{a_d} (weak anchoring) and large W_{a_0} (strong anchoring), $\phi(0)$ and $\phi(d)$ decreased with decreasing W_{a_d}, as shown in Figure 2b. The twist elastic torque in the bulk overcomes the surface anchoring torque. Thus, the LC director on the weak anchoring surface turns to the direction of the easy axis at the counter strong anchoring substrate [17–19]; $\phi(d)$ was 0 degrees when W_{a_d} was less than a certain value, which is called the threshold anchoring strength, a critical anchoring strength [17]. The critical anchoring strength was estimated to be 5.0×10^{-7} N/m (=K_{22}/d) with the infinitely strong W_{a_0} [17,19]. When W_{a_0} was set to be 1.0×10^{-4} N/m, which was the typical azimuthal anchoring strengths= of the rubbed PI alignment surface [7], the $\phi(d)$ curve almost overlapped with that estimated using infinite anchoring strengths (see also the extended figure shown in Figure 2b). To be precise, $\phi(d)$ was not zero but 4.9 degrees at W_{a_d} of 5.0×10^{-7} N/m. If W_{a_0} is set to be 1.0×10^{-5} N/m, the critical anchoring strength is estimated to be 4.8×10^{-7} N/m; $\phi(0)$ was 4.1 degrees at W_{a_d} of 1.0×10^{-5} (=W_{a_0}) N/m and it decreased to 0 degrees as well as $\phi(d)$ at the critical anchoring strength of 4.8×10^{-7} N/m. Then, the absolute homogeneous LC orientation, that is the Q-Homo-TN, was created by the asymmetric azimuthal anchoring strength.

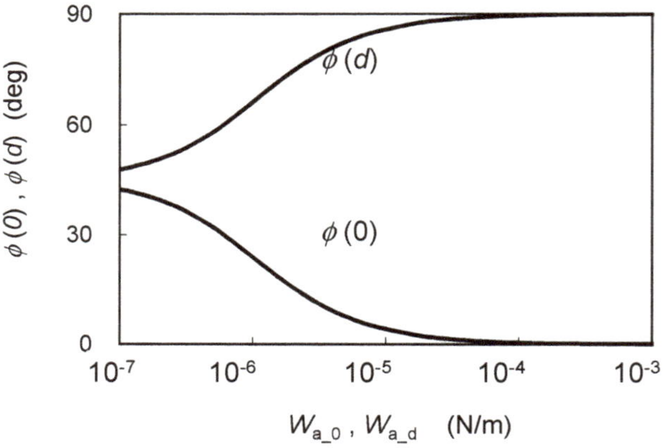

(a)

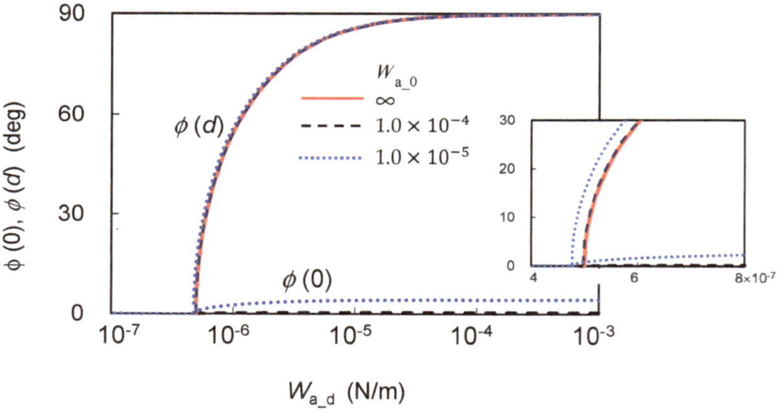

(b)

Figure 2. The $\phi(0)$ and $\phi(d)$ as a function of W_{p_d} in the TN cell with (**a**) symmetric and (**b**) asymmetric anchoring strengths of the alignment surfaces with the parameter of W_{a_0}.

We also calculated $\phi(d)$ in the case with non-zero pretilt angle surfaces. When the pretilt angle on both sides of the substrate was 2 degrees, calculated results with an extremely small difference were obtained as those shown in Figure 2a,b. As the pretilt angle increased more, we confirmed the increase in the twist angle between weak anchoring substrates. As the pretilt angle increased from 0 degrees to 10 degrees, the twist angle increased from 42.3 degrees to 43.1 degrees in the TN cell with symmetric azimuthal anchoring strength of 1.0×10^{-6} N/m, since the twist elastic torque of the LC orientation decreased with increasing the pretilt angle.

3. Electro-Optical Characteristics

3.1. Theoretical Electro-Optical Characteristics

The electro-optical property of the TN cell was estimated by using Jones matrix calculus at the wavelength of 550 nm. The cell was placed between crossed polarizers and the transmission axes of the polarizer were arranged with the easy axes of the alignment substrate. Figure 3 shows the electro-optical characteristic of the TN cell with the symmetric azimuthal anchoring strength. The polar anchoring of both sides of the substrate was 1.0×10^{-3} N/m. The tilt angle was set to be 2 degrees assuming the typical rubbed PI surface. The voltage–transmission (V-T) curve with the azimuthal anchoring strength of 1.0×10^{-4} N/m was almost the same as the V-T curve with using the infinitely strong anchoring. When the anchoring was 1.0×10^{-5} N/m, the twist angle in the absence of the applied voltage decreased to 81.8 degrees and the curve slightly shift to the right, which has already been reported by Inoue et al. [8]. The driving voltages V_{10}, that is the voltage necessary to obtain the transmittance of 10% in a "black on white mode", were 1.66 V and 1.71 V in the TN cells with the anchoring strength of 1.0×10^{-4} and 1.0×10^{-5} N/m, respectively. The twist angle without the voltage application decreased to 42.4 and 25.8 degrees when the anchoring became symmetrically weak to 1.0×10^{-6} and 5.0×10^{-7} N/m 1.0×10^{-5} N/m, respectively, and V-T curves were significantly changed from the normal V-T curve of the TN mode.

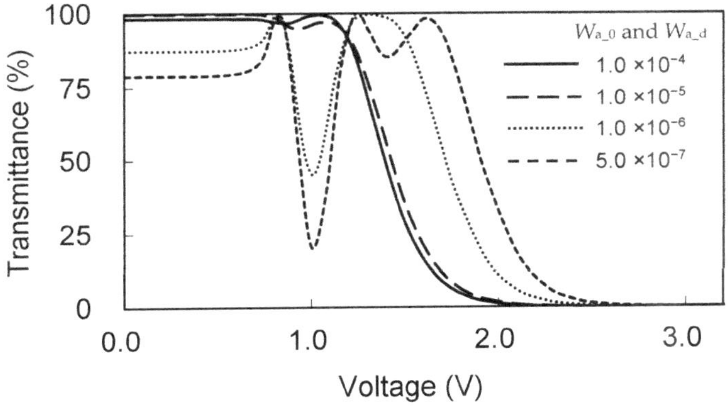

Figure 3. Voltage–transmittance curves of the TN cell with the symmetric azimuthal anchoring strength.

Figure 4 shows the tilt and twist angle distributions in the Q-Homo-TN cell with the asymmetric azimuthal anchoring strength under the voltage application. The azimuthal anchoring strengths of W_{a_0} and W_{a_d} were set to be 1.0×10^{-4} and 5.0×10^{-7} N/m, respectively. The θ_0 at the strong azimuthal anchoring surface was 2 degrees; θ_d at the weak azimuthal anchoring surface was 0 degrees assuming a very weak rubbed PI or a photoalignment surface. The polar anchoring of both sides of the substrate was 1.0×10^{-3} N/m. The LC molecules were homogeneously oriented in the voltage-off state. The threshold voltage at which the LC tilt angle at the center of the LC layer began to increase was about 0.4 V and was the same as that in the TN cell using the symmetric strong azimuthal anchoring strength of 1.0×10^{-4} N/m. The distributions of θ with the respective voltage application were also almost the same as that in the TN cell. On the other hand, $\phi(d)$ increased with the voltage in the Q-Homo-TN cell, because LC molecules glide on the weak anchoring surface due to the twist torque reduction in bulk [17]. When the applied voltage was 3 V, $\phi(d)$ reached 90 degrees and the ϕ distribution in the Q-Homo-TN cell was almost identical to that in the TN cell with the symmetric strong anchoring.

Figure 5 shows V-T curves of the TN cells with the parameter of W_{a_d}. W_{a_0} was set to be 1.0×10^{-4} N/m. The transmittance without the voltage decreased to 0% since the twist angle also decreased to 0 degrees with decreasing W_{a_d}. In the Q-Homo-TN cells with W_{a_d} of 5.0×10^{-7} N/m, θ and ϕ distribution shown in Figure 4 produced the V-T curve which exhibited 0–98.8–0% shift of the transmittance with the applied voltage. The V-T curve shows a display device operating in a "white on black mode" in the drive at the low-voltage side. The driving voltage of V_{90}, which is the voltage necessary to obtain the transmittance of 90%, was about 1.40 V, which shows the lower driving voltage of bright/dark switching compared to that in the TN cell. The maximum transmittance was obtained at 1.50 V and then decreased to 0% again because of the loss of optical rotation. When W_{a_d} was lower than the critical anchoring of 5.0×10^{-7} N/m, the threshold voltage and the driving voltage were increased, and the maximum transmittance decreased.

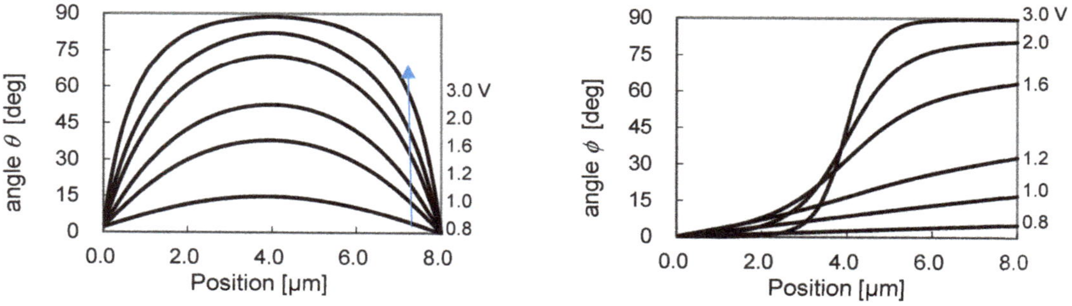

(a) (b)

Figure 4. Director distribution of (**a**) θ and (**b**) ϕ in the Q-Homo-TN cell with W_{a_0} of 1.0×10^{-4} N/m and W_{a_d} of 5.0×10^{-7} N/m.

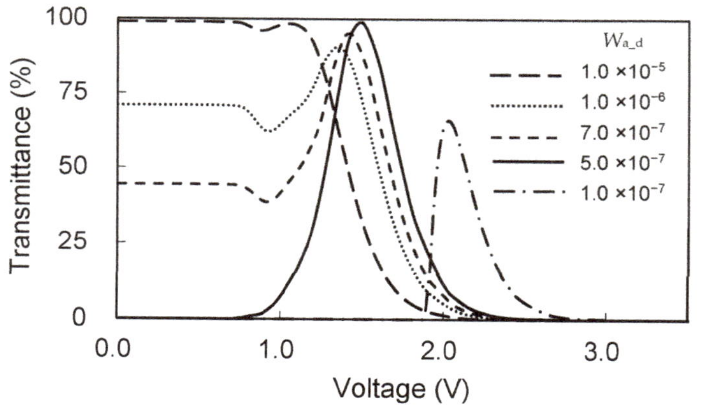

Figure 5. Voltage–transmittance curves of the TN cell with the asymmetric azimuthal anchoring strength.

V-T curves at 640 nm ($\Delta n = 0.170$), 550 nm ($\Delta n = 0.184$), and 460 nm ($\Delta n = 0.204$) corresponding to red, green, and blue wavelengths are shown in Figure 6. The wavelength dependence of the transmittance was small in the usual TN cell using strong anchoring substrates due to the contribution of optical rotation characteristics, as shown in Figure 6a. In the TN cell using symmetrically weak anchoring substrates, the wavelength dependence was large especially on the low voltage side by a birefringence effect of the LC, as shown in Figure 6b. The V-T curves of the red and green lights were very close, but not for the blue light in the Q-Homo-TN cell, as shown in Figure 6c. The incident light of 460 nm was close to the fourth Mauguin minimum conditions of the 90-degree TN cell [20]. The peak transmittance was improved from 83.5% to more than 90% by using the LC material with smaller Δn such that the second and third Mauguin minimum conditions are satisfied.

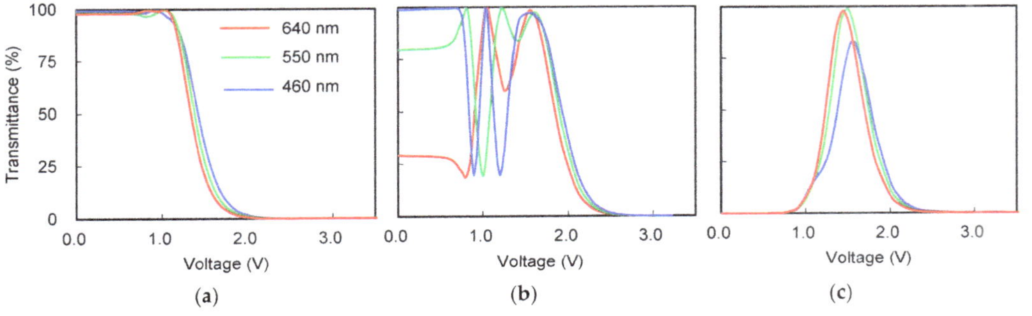

Figure 6. Voltage–transmittance curves at the wavelength of 640, 550, and 450 nm in the TN cell with the symmetric anchoring strength of (**a**) 1.0×10^{-4} N/m, (**b**) 5.0×10^{-6} N/m, and (**c**) in the Q-Homo-TN cell with the asymmetric anchoring strength of 1.0×10^{-4} and 5.0×10^{-6} N/m.

3.2. Cell Fabrication

It is known that the azimuthal anchoring strength of the PI film depends on the rubbing strength and the amount of UV irradiation in a photoalignment technology [21,22]. However, these surfaces have usually a strong surface memory effect even without the alignment treatment and have a certain strong azimuthal anchoring strength. The PVCi film is well known as the LC photoalignment fabricated by irradiating with a linearly polarized UV light [23–27]. The azimuthal anchoring of the PVCi surface increases to about 5.0×10^{-6} N/m with the UV dose [24,28], the strength of which is insufficient to achieve the twist angle of 90 degrees. In addition, it was reported that the anchoring strength increased to about 2.0×10^{-5} N/m when the rubbed PVCi film was exposed with linearly polarized UV light [29]. We also exposed the film with unpolarized UV light and obtained the same azimuthal anchoring strength of 2.0×10^{-5} N/m [19].

Figure 7a shows a schematic model of the fabricated LC cell with asymmetric azimuthal anchoring substrates using the rubbed PI (AL-1254 from JSR Corp. Japan) and PVCi surfaces. The cell thickness was controlled by using 9 μm ball spacers. The anchoring strength of the PVCi surface without the UV irradiation was about 8×10^{-8} N/m. We controlled the anchoring strength by the irradiation time of the UV light from a super-high-pressure Hg lamp source [19]. The UV power was 10 mW/cm^2. The left-half part of the rubbed PVCi surface was irradiated with the UV light for 20 s to obtain a target anchoring strength of 4.5×10^{-7} ($=K_{22}/d$) N/m. The right-half side was irradiated for 120 s and the anchoring strength was about 1.0×10^{-5} N/m. The easy axis of the rubbed PVCi surface is perpendicular to the rubbing direction [30–32]. Accordingly, 5CB was sandwiched between PVCi and PI substrates whose rubbing directions were parallel to each other. Figure 7b shows optical polarizing microscope images of the fabricated LC cell placed between crossed and parallel polarizers. The homogeneous orientation without the twist deformation and the optical rotation of roughly 90 degrees were confirmed on the weak and strong anchoring PVCi surfaces, respectively.

3.3. V-T Curves

Figure 8 shows the voltage–transmittance curves of the fabricated LC cell. The transmitted light intensity through parallel polarizers in the absence of the LC cell was defined as 100%. The wavelength of the incident light was 640 nm. The rate of the voltage sweep was 2 V/min. When the Q-Homo-TN cell was placed between crossed polarizers, the transmittance increased from 0% to 60% and then decreased to 0% again with increasing the voltage. This indicates that the in-plane rotation of the LC on the weak anchoring PVCi surface occurred, and the LC orientation changed from the homogeneous to TN orientation, as shown by the numerical results in Figure 4b. The hysteresis characteristics were observed with increasing and decreasing voltages. The in-plane switching mode cell using the PVCi surface has also been reported to exhibit a hysteresis characteristic [33,34]. It has also been reported that the high viscosity of the gliding LC molecules on the surface caused a very slow response [35–37]. Therefore, the hysteresis characteristic of the Q-Homo-TN cell might be caused by the slow response of the in-plane rotation of the LC on the PVCi surface. On the other hand, in the TN orientation between rubbed PI and fully crosslinked PVCi surfaces, a common V-T curve of the TN cell without hysteresis was obtained when the voltage sweep rate was 5 V/min.

(a)

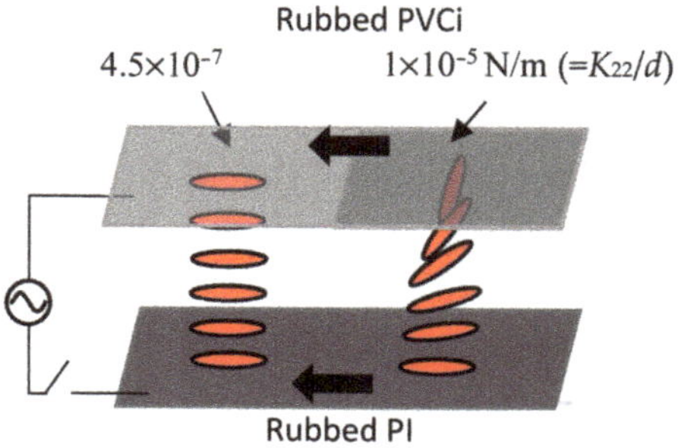

(b)

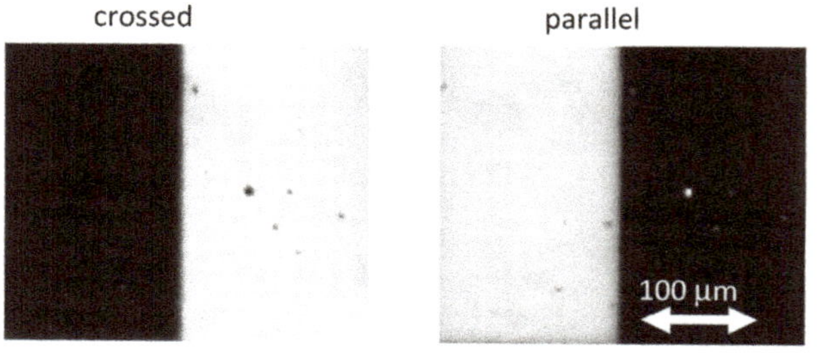

Figure 7. (**a**) Schematic model of the LC orientation of the Q-Homo-TN and TN between rubbed PVCi and PI substrates. (**b**) Optical polarizing microscope images of the LC cell placed between crossed and parallel polarizers.

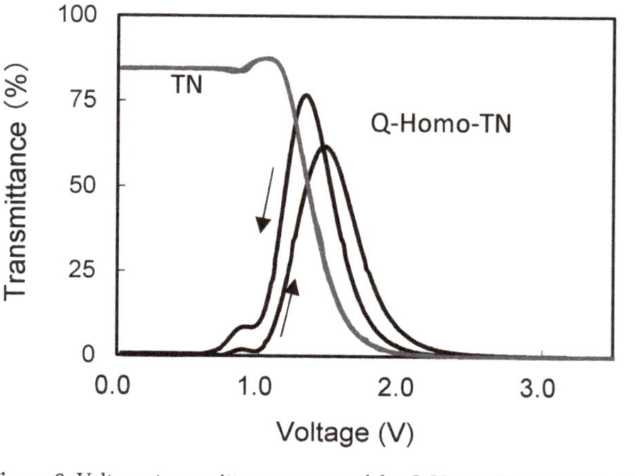

Figure 8. Voltage–transmittance curves of the Q-Homo-TN and TN cells at 640 nm.

We calculated the theoretical V-T curves using the following data: the polar and azimuthal anchoring strengths of the rubbed PI are 1.0×10^{-3} and 4.0×10^{-4} N/m, respectively. The pretilt angle is about 2 degrees. The polar anchoring strength of PVCi is 1.0×10^{-4} N/m [27]. The pretilt angle of the PVCi surface is 0 degrees since the LC aligns perpendicular to the rubbing direction. Converting the peak value of the measurement curve to 100%, the V-T curve of the fabricated TN cell was in close agreement with the calculated value at 640 nm. In the Q-Homo-TN cell, the measured peak transmittance was obtained at 1.46 V with the voltage increase, which almost coincides with the calculated peak voltage, even though the peak value in the fabricated cell was lower than the calculated peak value of 99.0%. Some alignment defects generated through the rubbing treatment on the uncrosslinked PVCi surface resulted in a lower peak value of the transmittance. In addition, the transmittance became maximum at 1.33 V and the peak value was higher than that increasing the voltage. The calculated results of the director distributions shown in Figure 4 were in the steady-state liquid crystal reorientation states, and the hysteresis properties as reported in highly twisted nematic cells [38,39] were not confirmed in our calculation.

Another problem with using the weak anchoring surface has been reported as an alignment memory effect [12,40]. Some kinds of alignment film have been proposed as the weak anchoring surface without the memory effect, for example a para-PVCN-F [12] surface and a high-density, concentrated brush poly (n-hexyl methacrylate) grafted substrate [13,14]. These alignment films and further studies on relations between the weak anchoring surface and the liquid crystal interface will lead to better performance of this device.

4. Conclusions

The electro-optical characteristics in the TN cell with symmetric and asymmetric anchoring strength were investigated. The Q-Homo-TN orientation was successfully fabricated using the rubbed PI surface and the PVCi surface, in which azimuthal anchoring strength was adjusted to the critical value. The dark–bright–dark switching of the Q-Homo TN cell was observed in the crossed polarizers when applying the voltage, which indicates the 90-degree in-plane rotation of LC molecules by the electric field perpendicular to the substrate. The hysteresis characteristics of the V-T curve were observed in the fabricated Q-Homo-TN cell, which was not shown in the claustration. In future work, we intend to further investigate response and decay times at various anchoring strengths by changing the crosslinking degree.

Author Contributions: Conceptualization, R.Y.; Methodology, cell fabrication, and evaluation, R.Y. and Y.S.; Writing—original draft preparation, R.Y.; Writing—review and editing, R.Y. and Y.S. All authors have read and agreed to the published version of the manuscript.

Funding: This research received no external funding.

Data Availability Statement: Not applicable.

Conflicts of Interest: The authors declare no conflict of interest.

References

1. Schadt, M.; Helfrich, W. Voltage-dependent optically activity of a twisted nematic liquid crystal. *Appl. Phys. Lett.* **1971**, *18*, 127–128. [CrossRef]
2. de Gennes, P.G. *The Physics of Liquid Crystals*; Clarendon: Oxford, UK, 1974; pp. 59–79.
3. Rapini, A.; Papoular, M. Distorsion d'une lamelle nematique sous champ magnetique conditions d'ancrage aux parois. *J. Phys. (Paris) Colloq.* **1969**, *30*, C4-54–C4-56. [CrossRef]
4. Nehring, J.; Kmetz, A.R.; Scheffer, T.J. Analysis of weak-boundary-coupling effects in liquid-crystal displays. *J. Appl. Phys.* **1976**, *47*, 850–857. [CrossRef]
5. Sprang, H.A.; Breddels, P.A. Numerical calculations of director patterns in highly twisted nematic configurations with nonzero pretilt angles. *J. Appl. Phys.* **1986**, *60*, 968–972. [CrossRef]
6. Hirning, R.; Funk, W.; Trebin, H.-R.; Schmidt, M.; Schmiedel, H. Threshold behavior and electro-optical properties of twisted nematic layers with weak anchoring in the tilt and twist angle. *J. Appl. Phys.* **1991**, *70*, 4211–4216. [CrossRef]

7. Ishihara, S.; Mizusaki, M. Alignment control technology of liquid crystal molecules. *J. Soc. Inf. Disp.* **2019**, *28*, 44–74. [CrossRef]
8. Inoue, M.; Manabe, N.; Akahane, T. Influence of the azimuth anchoring energy on the electro-optical characteristics of LCDs. In Proceedings of the 5th International Display Workshop, LCTp2-1. Kobe, Japan, 7–9 December 1998; pp. 41–44.
9. Oh-e, M.; Kondo, K. Electro-optical characteristics and switching behavior of the in-plane switching mode. *Appl. Phys. Lett.* **1995**, *67*, 3895–3897. [CrossRef]
10. Yoneya, M.; Iwasaki, K.; Tomioka, Y.; Kondo, K. Cell gap margin enlargement of in-plane switching mode liquid crystal displays using weak-anchoring effects. *Appl. Phys. Lett.* **1999**, *74*, 803–805. [CrossRef]
11. Yoneya, M.; Kondo, K. Threshold behavior analysis of in-plane switching mode liquid-crystal cells with asymmetrical surface condition. *Appl. Phys. Lett.* **1999**, *74*, 3477–3479. [CrossRef]
12. Andrienko, D.; Barbet, F.; Bormann, D.; Kurioz, Y.; Kwon, S.-B.; Reznikov, Y.; Warenghem, M. Electrically controlled director slippage over a photosensitive aligning surface; in-plane sliding mode. *Liq. Cryst.* **2000**, *27*, 365–370. [CrossRef]
13. Sato, O.; Iwata, N.; Kawamura, J.; Maeda, T.; Tsujii, Y.; Watanabe, J.; Tokita, M. An in-plane switching liquid crystal cell with weakly anchored liquid crystals on the electrode substrate. *J. Mater. Chem. C* **2017**, *5*, 4384–4387. [CrossRef]
14. Sato, O.; Okuno, H.; Adachi, I.; Goto, K.; Noda, T.; Tsutsui, K. A high transmittance and fast response in-plane switching liquid crystal display with the zero-azimuth anchoring layers on the electrodes. *J. Phys. D Appl. Phys.* **2020**, *53*, 15LT02. [CrossRef]
15. Lee, S.H.; Lee, S.L.; Kim, H.Y. Electro-optic characteristics and switching principle of a nematic liquid crystal cell controlled by fringe-field switching. *Appl. Phys. Lett.* **1998**, *73*, 2881–2883. [CrossRef]
16. Choi, Y.; Oh, S.-W.; Choi, T.-H.; Sohn, H.-J.; Do, S.-M.; Yoon, T.-H. Liquid crystal cell asymmetrically anchored for high transmittance and triggered with a vertical field for fast switching. *Opt. Express* **2020**, *28*, 20553–20562. [CrossRef] [PubMed]
17. Yamaguchi, R. Analysis of Electro-Optical Behavior in Liquid Crystal Cells with Asymmetric Anchoring Strength. *Symmetry* **2022**, *14*, 85. [CrossRef]
18. Strigazzi, A. On the critical thickness of a twisted nematic cell. *J. Phys.* **1985**, *46*, 1507–1512. [CrossRef]
19. Yamaguchi, R.; Sato, S. Continuous grey scale image printing on the liquid crystal cell. *Appl. Phys. Lett.* **2005**, *86*, 031913-1–031913-3. [CrossRef]
20. Gooch, C.H.; Tarry, H.A. The optical properties of twisted nematic liquid crystal structures with twist angles $\leq$90 degrees. *J. Phys. D Appl. Phys.* **1975**, *8*, 1575–1584. [CrossRef]
21. Seo, D.-S. Relationship between surface anchoring strength and surface ordering on weakly rubbed polyimide surfaces. *Liq. Cryst.* **2001**, *27*, 1539–1542. [CrossRef]
22. Hasegawa, M. Modeling of photoinduced optical anisotropy and anchoring energy of polyimide exposed to linearly polarized deep UV Light. *Jpn. J. Appl. Phys.* **1999**, *38*, L457–L460. [CrossRef]
23. Schadt, M.; Schmitt, K.; Kozinkov, V.; Chigrinov, V. Surface-induced parallel alignment of liquid crystals by linearly polymerized photopolymers. *Jpn. J. Appl. Phys.* **1992**, *31*, 2155–2164. [CrossRef]
24. Bryan-Brown, G.P.; Sage, I.C. Photoinduced ordering and alignment properties of polyvinylcinnamates. *Liq. Cryst.* **1996**, *20*, 825–829. [CrossRef]
25. Chen, J.; Johnson, D.L.; Bos, P.J.; Wang, X.; West, J.L. Model of liquid crystal alignment by exposure to linearly polarized ultraviolet light. *Phys. Rev. E* **1996**, *54*, 1599–1603. [CrossRef]
26. Ichimura, K.; Akita, Y.; Akiyama, H.; Kudo, K.; Hayashi, Y. Photoreactivity of polymers with regioisomeric cinnamate side chains and their ability to regulate liquid crystal alignment. *Macromolecules* **1997**, *30*, 903–911. [CrossRef]
27. Iimura, Y.; Kobayashi, S.; Hashimoto, T.; Sugiyama, T.; Katoh, K. Alignment control of liquid crystal molecules using photodimerization reaction of poly (vinyl cinnamate). *IEICE Trans. Electron.* **1996**, *E79-C*, 1040–1046.
28. Vilfan, M.; Olenik, I.D.; Mertelj, A.; Copic, M. Aging of surface anchoring and surface viscosity of a nematic liquid crystal on photoaligning poly-(vinyl-cinnamate). *Phys. Rev. E* **2001**, *63*, 061709-1–061709-5. [CrossRef] [PubMed]
29. Li, X.T.; Saitoh, H.; Nakamura, H.; Kobayashi, S.; Iimura, Y. Mechanism of photo-induced liquid crystal alignment on polyvinyl cinnamate surface. *J. Photopolym. Sci. Technol.* **1997**, *10*, 13–17. [CrossRef]
30. Yamaguchi, R.; Goto, Y.; Sato, S. A novel patterning method of liquid crystal alignment by azimuthal anchoring control. *Jpn. J. Appl. Phys.* **2002**, *41*, L889–L891. [CrossRef]
31. Yamaguchi, R.; Sato, S. Liquid crystal material dependence on rubbed PVCi alignment properties. *Mol. Cryst. Liq. Cryst.* **2010**, *516*, 32–37. [CrossRef]
32. Yamaguchi, R.; Nishimura, M.; Ikeya, M. Azimuthal anchoring change by liquid crystal mixtures on poly (vinyl Cinnamate) film. *J. Photopolym. Sci. Technol.* **2013**, *26*, 393–396. [CrossRef]
33. Zhou, Y.; Sato, S. Electrooptical and response/relaxation properties of liquid crystal cells in in-plane switching mode with polyvinylcinnamate photoinduced alignment layer. *Jpn. J. Appl. Phys.* **1998**, *37*, 4439–4443. [CrossRef]
34. Li, X.T.; Kawakami, A.; Akiyama, H.; Kobayashi, S.; Iimura, Y. Reduction in driving voltage of in-plane switching liquid crystal displays using photo-alignment method. *Jpn. J. Appl. Phys.* **1998**, *37*, L743–L745. [CrossRef]
35. Kurioz, Y.; Reshetniak, V.; Reznikov, Y. Orientation of a liquid crystal on a soft photoaligning surface. *Mol. Cryst. Liq. Cryst.* **2002**, *375*, 535–541. [CrossRef]
36. Buluy, O.; Iljin, A.; Ouskova, E.; Reznikov, Y.; Blanc, C.; Nobili, M.; Antonova, K. Anchoring and gliding of easy axis of 5CB on photoaligning PVCN-F surface. *J. Soc. Inf. Disp.* **2006**, *14*, 603–610. [CrossRef]

37. Mema, E.; Kondic, L.; Cummings, L.J. Director gliding in a nematic liquid crystal layer: Quantitative comparison with experiments. *Phys. Rev. E* **2018**, *97*, 032704. [CrossRef]
38. Saito, S.; Kamihara, M.; Kobayashi, S. Influence on the hysteresis effect of various parameters in supertwisted nematic liquid crystals. *Mol. Cryst. Liq. Cryst.* **1986**, *139*, 171–187. [CrossRef]
39. Li, J.; Hoke, C.D.; Bos, P.J. Studies of the Bistability of Highly Twisted Nematics. *Jpn. J. Appl. Phys.* **1996**, *35*, L706. [CrossRef]
40. Prakash, J.; Choudhary, A.; Kaur, S.; Mehta, D.S.; Biradar, A.M. Memory effect in weakly anchored surfaces of deformed helix ferroelectric liquid crystals. *Phys. Rev. E* **2008**, *78*, 021707. [CrossRef] [PubMed]

symmetry

Article

Analysis of Electro-Optical Behavior in Liquid Crystal Cells with Asymmetric Anchoring Strength

Rumiko Yamaguchi

Electrical and Electronic Engineering Course, Graduate School of Engineering Science, Akita University, Akita 010-8502, Japan; yrumiko@gipc.akita-u.ac.jp; Tel.: +81-18-889-2483

Abstract: Liquid crystal director distributions have been numerically analyzed between asymmetric anchoring surfaces, that is, infinitely strong and very weak anchoring strength interfaces. In a hybrid aligned nematic (HAN) cell and a twisted nematic (TN) cell, HAN and TN orientations turn to a homogeneous orientation when the weak anchoring strength is lower than a critical one. Relationships between the anchoring strength and elastic constants of the liquid crystal were analyzed to be of a quasi-homogeneous orientation. The quasi-homogeneous orientation returned to the original HAN and TN orientations under voltage application. Low-driving electro-optical properties with no threshold voltage can be obtained in a quasi-homogeneous HAN cell. A unique voltage–transmission curve of 0–100–0% appeared in a quasi-homogeneous TN cell between the crossed polarizers.

Keywords: nematic liquid crystal; polar anchoring; azimuthal anchoring; hybrid aligned nematic; twisted nematic; homogeneous orientation; threshold voltage; transmittance

Citation: Yamaguchi, R. Analysis of Electro-Optical Behavior in Liquid Crystal Cells with Asymmetric Anchoring Strength. *Symmetry* **2022**, *14*, 85. https://doi.org/10.3390/sym14010085

Academic Editors: Shoichi Ishihara and Sadahito Uto

Received: 4 December 2021
Accepted: 2 January 2022
Published: 6 January 2022

1. Introduction

Nematic liquid crystals (NLCs) are widely applied to information displays and optical devices of adaptive lenses, polarization grating, spatial light modulators, etc. [1]. The NLC is usually sandwiched between two glass substrates, the surfaces of which are covered with an alignment film. The alignment film surface has an easy axis and controls the NLC's orientation in the bulk of the LC cell. Conventional NLC displays are fabricated using a strong anchoring alignment surface on both sides of the substrate to maintain a specific NLC orientation. A Fréedericksz transition is one of the most famous electro-optical switching mechanisms for NLCs [2], and a typical threshold voltage is approximately 1–2 volts between strong anchoring alignment surfaces. Therefore, the driving voltage for an LC display is three times the threshold voltage or more in a typically twisted nematic (TN) display mode.

On the other hand, some types of LC displays using asymmetric anchoring strength surfaces, that is, infinitely strong and very weak anchoring alignment surfaces, have been proposed to reduce the driving voltage [3–5]. Interdigital electrodes have been fabricated on weak azimuthal anchoring substrates. A homogeneous (or TN) LC orientation in an initial state changes to a TN (or homogeneous) orientation by in-plane field, since the LC glides on the weak anchoring surface [6–8]. A polarization rotator has also been reported by using asymmetric azimuthal anchoring surfaces. LC molecules glide to the in-plane field direction on the weak anchoring surface, and LCs on the counter-strong anchoring substrate are fixed. Then, the twist angle changes, which rotates the polarization direction of the incident light [9].

Another interest for asymmetric anchoring surfaces with strong and weak anchoring strengths is the competition between the anchoring effect and elastic torque. Although the LC director prefers to align to the easy axis with the alignment surface, the actual director at the weak surface may glide from the easy axis to minimize the free energy of the LC cell, even if without an external field. The asymmetric polar anchoring strength has been theoretically investigated in a hybrid aligned nematic (HAN) cell. The HAN orientation changes to the homogeneous or homeotropic orientation when the thickness of the LC layer decreases to a critical thickness [10–12]. Papers have discussed the tilt angle on the weak anchoring substrate as a function of the external electric field or magnetic field. A flexo-electro-optic effect has also been investigated theoretically and experimentally in a HAN cell with asymmetric polar anchorings [13]. Moreover, Strigazzi analyzed the effect of weak azimuthal anchoring in a TN cell and the existence of the critical thickness limiting the TN orientation construction [14].

In this paper, critical anchoring is theoretically discussed instead of the critical thickness in HAN and TN cells. The LC director distributions and electro-optical properties were numerically estimated as a function of the applied voltage. A threshold-less electro-optical curve and a low-driving voltage voltage–transmission curve were obtained.

2. Principle

2.1. Free Energy of LC Cells

Figure 1 shows the definition of the liquid crystal director, n, with a tilt angle, θ, and a twist angle, φ. W_{p_0} and W_{p_d} are the polar anchoring strengths, and W_{a_0} and W_{a_d} are azimuthal anchoring strengths of the lower and upper substrates, respectively.

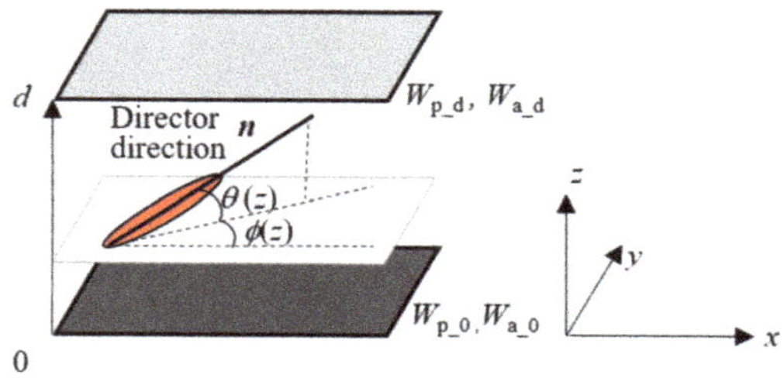

Figure 1. Definition of the liquid crystal director, n, and anchoring strength, W.

When W_{p_0} and W_{a_0} of the lower side of the substrate are strong (infinite), the total free energy per unit area, F, in the LC cell [10,13,14] is given by:

$$
\begin{aligned}
F &= F_{\text{bulk}} + F_{\text{electric}} + F_{\text{surface}} \\
F_{\text{bulk}} + F_{\text{electric}} &= \int_0^d \frac{1}{2}\Big\{ \left(K_{11}\cos^2\theta(z) + K_{33}\sin^2\theta(z)\right)\left(\tfrac{d\theta(z)}{dz}\right)^2 \\
&\quad + \left(K_{22}\cos^2\theta(z) + K_{33}\sin^2\theta(z)\right)\cos^2\theta(z)\left(\tfrac{d\phi(z)}{dz}\right)^2, -\varepsilon_0\left(\varepsilon_\perp + \Delta\varepsilon sin^2\theta(z)\right)\left(\tfrac{dV(z)}{dz}\right)^2 \Big\} \\
F_{\text{surface}} &= \tfrac{1}{2}W_{p\_d}\sin^2(\theta_d - \theta(d)) + \tfrac{1}{2}W_{a\_d}\sin^2(\phi_d - \phi(d)),
\end{aligned}
\tag{1}
$$

where K_{11}, K_{22}, and K_{33} are the splay, twist, and bend elastic constants; d is the thickness of the LC layer; $V(z)$ is the voltage potential. θ_d and φ_d are easy axis angles of the upper side of the substrate. $\theta(0)$ and $\phi(0)$ at the lower side of the substrate are fixed at 0.

In the case of the HAN cell, θ_d is $\frac{\pi}{2}$, and φ_d is 0. Without the field, the Euler–Lagrange equation leads to the splay-bend torque balance equation [13] at the homeotropic surface as follows:

$$
\left(K_{11}\cos^2\theta(d) + K_{33}\sin^2\theta(d)\right)\frac{\theta(d)}{d} = W_{p\_d}\sin\left(\frac{\pi}{2} - \theta(d)\right)\cos\left(\frac{\pi}{2} - \theta(d)\right). \tag{2}
$$

In the TN cell, θ_d is 0 and φ_d is $\frac{\pi}{2}$. Therefore, the twist torque balance equation [9,15] is given by:

$$
K_{22}\frac{\varphi(d)}{d} = W_{a\_d}\sin\left(\frac{\pi}{2} - \varphi(d)\right)\cos\left(\frac{\pi}{2} - \varphi(d)\right). \tag{3}
$$

2.2. Critical Anchoring

In the case of the asymmetrical anchorage effect, the upper substrate was so weak that the LC on the upper substrate could not keep the splay-bend and twist elastic torque in bulk. Consequently, the HAN and TN orientations turned to the homogeneous orientation without the orientation distortion as shown in Figure 2. The homogeneous orientation, shown in Figure 2c, may be termed as quasi-homogeneous (Q-Homo) orientation.

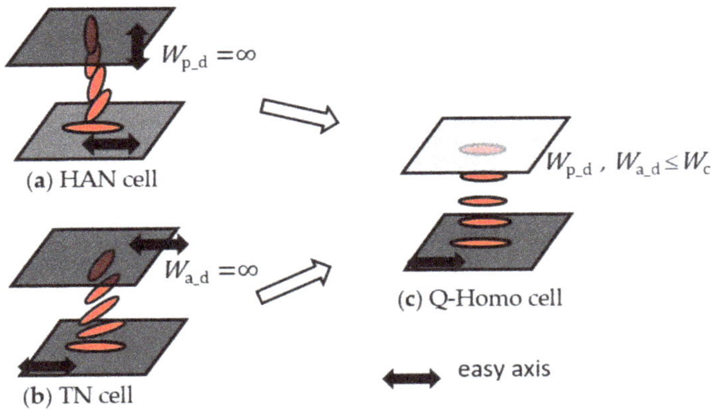

Figure 2. Schematic models of the (**a**) HAN, (**b**) TN, and (**c**) Q-Homo cells.

Table 1 shows the physical parameters of the LC used in the numerical calculations, which were typical values from a practical point of view. $\theta(d)$ was estimated as a function of the homeotropic polar anchoring, W_{p_d}, using Equation (2) in the HAN cell as shown in Figure 3a. The cell thickness, d, was 8 μm. $\theta(d)$ decreased with the decrease in the anchoring strength and was zero at a W_{p_d} of 1.5×10^{-6} N/m. Such a critical value was estimated as K_{11}/d and was independent of K_{33} as also shown in Figure 3. The twist angle φ (d) also decreased with decreasing W_{a_d} in the TN cell as shown in Figure 3b. The critical anchoring was estimated as K_{22}/d ($=1.0 \times 10^{-6}$ N/m) using Equation (3). The change in the twist angle was experimentally confirmed by controlling the anchoring of a poly(vinyl cinnamate) film [15].

Table 1. Physical parameters of the LC.

K_{11}	K_{22}	K_{33} (pN)	$\varepsilon_{//}$	$\varepsilon_{//}$	n_o	n_e
12	8	18	15	5	1.5	1.7

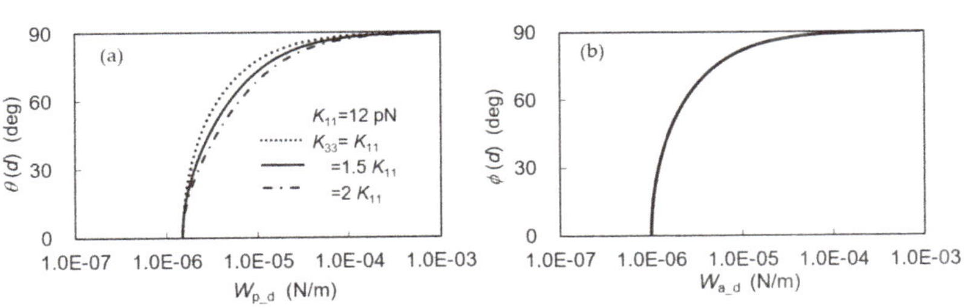

Figure 3. (**a**) W_{p_d} vs. tilt angle $\theta(d)$ in the HAN cell and (**b**) W_{a_d} vs. twist angle $\varphi(d)$ in the TN cell.F.

3. Electro-Optical Property in the Q-Homo Cell

3.1. Q-Homo–HAN Cell

Figure 4 shows $\theta(d)$ as a function of the applied voltage in the HAN cell with the parameter of W_{p_d}. The LC's physical properties, as shown in Table 1, were used in this section. $\theta(d)$ increased with the applied voltage and reached 90° at an applied voltage higher than 3 V. In the Q-Homo state of the HAN (Q-Homo-HAN) cell with an anchoring of 1.5×10^{-6} N/m (=W_c), $\theta(d)$ increased from 0° without the threshold voltage. The threshold voltage, V_{th}, appeared when W_{p_d} was less than W_c. V_{th} was 0.48 V in the cell when W_{p_d} was $0.1W_c = 1.5 \times 10^{-6}$ N/m.

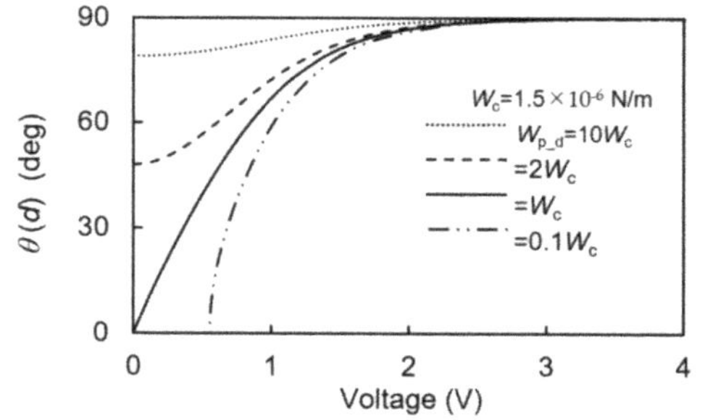

Figure 4. $\theta(d)$ as a function of the applied voltage.

Figure 5a shows the LC director distributions of θ in the Q-Homo-HAN cell with the parameter of the applied voltage. W_{p_d} was 1.5×10^{-6} N/m (=W_c). The LC tilt angle at the center of the cell was 30° by applying a voltage of only 0.5 V. Moreover, the θ distribution at 4 V was almost the same as that in the conventional HAN cell with infinite anchoring on both sides of the substrate. From these curves, an effective extraordinary refractive index $< n_e >$ was estimated as follows:

$$< n_e >= \frac{1}{d}\int_0^d \frac{n_e n_o}{\sqrt{\cos^2\theta(z){n_o}^2 + \sin^2\theta(z){n_e}^2}}dz, \tag{4}$$

where n_o and n_e are ordinary and extraordinary indices, respectively. Figure 5b shows $< n_e >$ vs. applied voltage in Q-Homo-HAN cells with a W_{p_d} of W_c and 0.1 W_c. Conventional Homo and HAN cells are also shown in Figure 5a. The Homo cell had a threshold voltage of approximately 1.16 V when using the LC parameters shown in Table 1. In the Q-Homo-HAN cell with W_c, $< n_e >$ decreased without the threshold voltage and was approximately 1.53 at 2 V, which was almost the same value as the HAN cell. Therefore, a large index change of approximately 0.17 was obtained in the Q-Homo-HAN cell compared to 0.06 in the HAN cell and 0.10 in the Homo cell at 2 V.

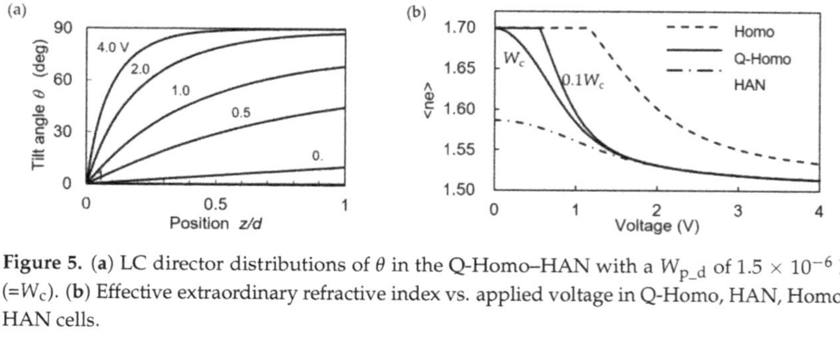

Figure 5. (**a**) LC director distributions of θ in the Q-Homo–HAN with a W_{p_d} of 1.5×10^{-6} N/m (=W_c). (**b**) Effective extraordinary refractive index vs. applied voltage in Q-Homo, HAN, Homo, and HAN cells.

A guest–host (GH) mode of the electro-optical property in the Q-Homo–HAN cell was also compared with the HAN and Homo cells. Typical absorption coefficients of 0.03 and 0.3 μm^{-1} for short and long axes of a dichroic dye were used, respectively [16]. The incident light was polarized parallel to the LC director of the planar aligned surface. In this case, the maximum (homeotropic orientation) and the minimum (homogeneous orientation) transmittances were estimated at 78.7 and 9.1%, respectively. Figure 6 shows transmittance vs. applied voltage curves in three cells. In the HAN cell, the transmittance in the voltage on state was higher than that in the Homo cell. However, the transmittance in the off state was also high, and the contrast ratio was very low. The transmittance in the on state of the Homo cell was lower, since the LCs did not reorient on both the strong anchoring planar surfaces. The driving voltage, V_{90}, at which the transmittance increased to T_{90} (=~72%) was approximately 13 V. On the other hand, V_{90} could decrease to 7 V in the Q-Homo-HAN cell.

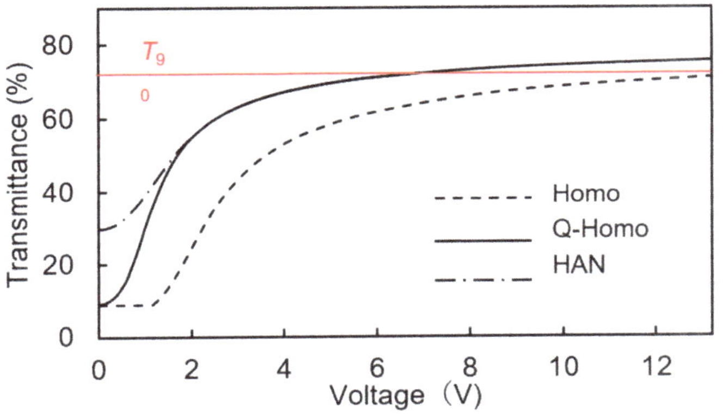

Figure 6. Transmittance as a function of the applied voltage in the GH mode.

3.2. Q-Homo-TN Cell

Figure 7 shows $\varphi(d)$ as a function of the applied voltage in the TN cell with the parameter of W_{a_d}. Elastic constants and dielectric constants shown in Table 1 were used in this section. The polar anchoring of both sides of the substrate was infinite. $\varphi(d)$ on the weak azimuthal anchoring surface increased with the applied voltage, since the twist torque in bulk decreased with the reorientation of the LC perpendicular to the substrate. Therefore, curves had a threshold voltage which corresponded to the voltage when the tilt angle started to increase. V_{th} was 1.16 V in the TN cell as well as in the homogeneous cell when using the cell parameter shown in Table 1. When W_{a_d} was less than W_c, for example, 0.5 W_c, V_{th} increased to approximately 2.4 V. $\varphi(d)$ turned to the original easy axis φ_d of $\frac{\pi}{2}$ by applying a voltage of more than 5 V. Figure 8 shows the schematic models of LC director distribution when the voltage was applied. Its starts from the homogeneous orientation to the TN orientation, and I call this type of LC cell a Q-Homo-TN cell.

Tilt and twist angle distributions were numerically estimated with the applied voltages and are shown in Figure 9. The polar anchoring strength was infinite and the pretilt angles θ (0) and θ (d) were 0°. The distributions of θ with the respective applied voltages were almost the same as those in a conventional TN cell using both strong anchoring surfaces. On the other hand, the φ distribution changed from the homogeneous state to the TN state. Both the θ and φ distributions were exactly the same as those in the conventional TN at 5.0 V.

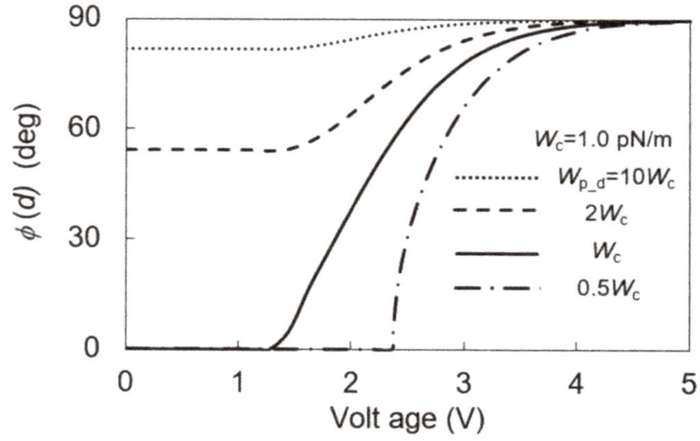

Figure 7. $\varphi(d)$ as a function of the applied voltage.

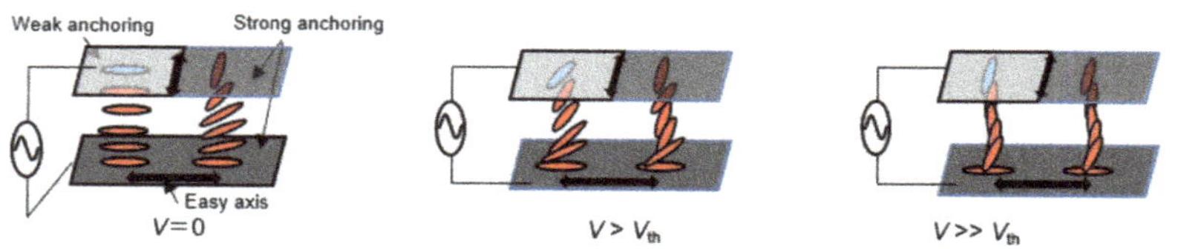

Figure 8. Schematic model of LC director distribution under the respective applied voltages in Q-Homo–TN cell (**left side**) and the conventional TN (**right side**) cell.

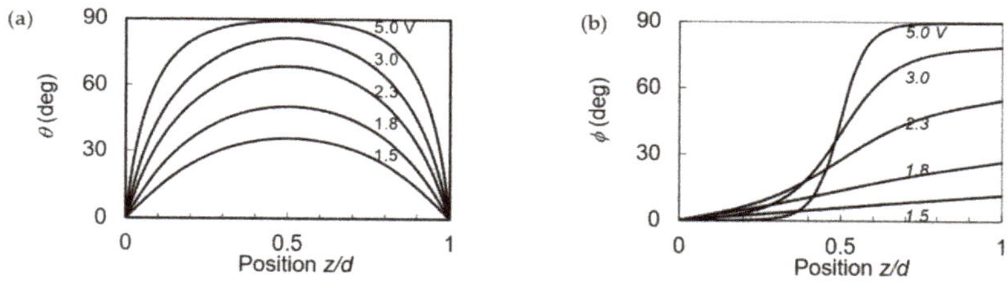

Figure 9. Director distributions of (**a**) θ and (**b**) φ.

Subsequently, the voltage–transmittance curves of the cell between the crossed polarizers were estimated using Jones matrix calculus [17,18] as shown in Figure 10a. The wavelength was 550 nm, which is very close to the third minimum condition; therefore, the transmittance in the voltage off state was 100% in the 90° TN cell. When the voltage was applied, the LC director reoriented along the electric field as shown in Figure 9a, and the twist angle distribution also changed. Thus, the polarization guiding effect was destroyed, and the transmittance reduced to 0%. On the other hand, in the Q-Homo-TN cell, it increased from 0 to 98.4% at 2.35 V, which was approximately two times that of the threshold voltage. Then, it decreased to 0% again because of the same circumstances of the TN cell. IF Δn is 0.18, 100% transmittance can be obtained at 2.31 V. The polarization states of the transmission light with the respective applied voltages are also shown in Figure 10b. When the lineally polarized light parallel to the x-axis was incident on the cell, the major axis of the elliptical polarized transmission light rotated to the y-axis when the voltage increased to 2.3 V and then returned to the x-axis again when the voltage increased to 4.0 V.

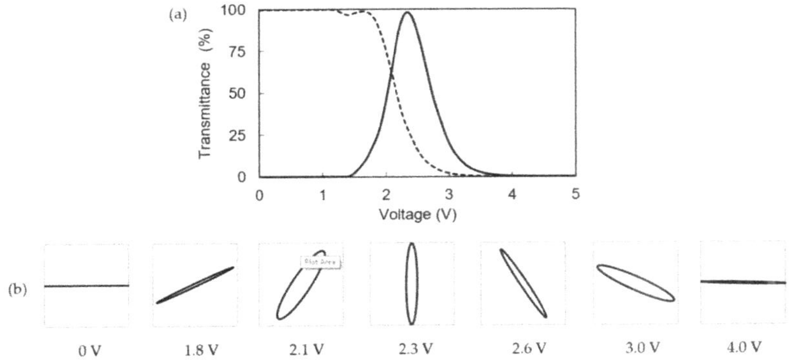

Figure 10. (**a**) Voltage–transmittance curves in Q-Homo–TN and TN cells. (**b**) Polarization states of the transmission light in the Q-Homo–TN cell with the respective applied voltages.

4. Conclusions

Electro-optical properties were numerically analyzed in liquid crystal cells with asymmetric anchoring strengths. The hybrid orientation turned to the homogeneous orientation when the polar anchoring of the homeotropic alignment surface reduced to the critical anchoring of K_{11}/d. The quasi-homogeneous HAN cell had no threshold voltage. Larger optical variations could be obtained by using a lower applied voltage compared to the conventional homogeneous cell. The twisted orientation turned to the homogeneous orientation when the azimuthal anchoring of the one side of the planar alignment surface reduced to the critical anchoring of K_{22}/d. The transmittance of the quasi-homogeneous TN cell between the crossed polarizers increased from 0% to approximately 100% at a voltage of approximately $2V_{th}$. Liquid crystal display modes with low-driving voltages can be clarified by using asymmetric anchoring strength.

Funding: This research received no external funding.

Institutional Review Board Statement: Not applicable.

Informed Consent Statement: Not applicable.

Data Availability Statement: Not applicable.

Conflicts of Interest: The author declares no conflict of interest.

References

1. Jaroszewicz, L.; Bennis, N. *Liquid Crystal Optical Device*; MDPI: Basel, Switzerland, 2020. [CrossRef]
2. Frederick, V.K.; Zolina, V. Forces causing the orientation of an anisotropic liquid. *Trans. Faraday Soc.* **1933**, *29*, 919–930. [CrossRef]
3. Yoneya, M.; Iwasaki, K.; Tomioka, Y.; Kondo, K. Cell gap margin enlargement of in-plane switching mode liquid crystal displays using weak-anchoring effects. *Appl. Phys. Lett.* **1999**, *74*, 803–805. [CrossRef]
4. Bryan-Brown, G.P.; Wood, E.L.; Sage, I.C. Weak surface anchoring of liquid crystals. *Nature* **1999**, *399*, 338–340. [CrossRef]
5. Andrienko, D.; Barbet, F.; Bormann, D.; Kurioz, Y.; Kwon, S.-B.; Reznikov, Y.; Warenghem, M. Electrically controlled director slippage over a photosensitive aligning surface; in-plane sliding mode. *Liq. Crysts.* **2000**, *27*, 365–370. [CrossRef]
6. Sato, O.; Iwata, N.; Kawamura, J.; Maeda, T.; Tsujii, Y.; Watanabe, J.; Tokita, M. An in-plane switching liquid crystal cell with weakly anchored liquid crystals on the electrode substrate. *J. Mater. Chem. C* **2017**, *5*, 4384–4387. [CrossRef]
7. Choi, Y.; Oh, S.-W.; Choi, T.-H.; Sohn, H.-J.; Do, S.-M.; Yoon, T.-H. Liquid crystal cell asymmetrically anchored for high transmittance and triggered with a vertical field for fast switching. *Opt. Express* **2020**, *28*, 20553–20562. [CrossRef] [PubMed]
8. Sato, O.; Okuno, H.; Adachi, I.; Goto, K.; Noda, T.; Tsutsui, K. A high transmittance and fast response in-plane switching liquid crystal display with the zero-azimuth anchoring layers on the electrodes. *J. Phys. D Appl. Phys.* **2020**, *53*, 15LT02. [CrossRef]
9. Yamaguchi, R.; Yamanaka, T.; Sato, S. Liquid-Crystal Optical Rotator Using Weak Azimuthal Anchoring Surface. *Jpn. J. Appl. Phys.* **2001**, *40*, 6522–6525. [CrossRef]
10. Barbero, G.; Barberi, R. Critical thickness of a hybrid aligned nematic liquid crystal cell. *J. Phys.* **1983**, *44*, 609–616. [CrossRef]
11. Barbero, G.; Bartolino, R. Remarks on critical thickness of a hybrid aligned nematic cell with large elastic anisotropy. *Mol. Cryst. Liq. Cryst.* **1983**, *99*, 89–97. [CrossRef]

12. Strigazzi, A. Second order elasticity and critical thickness of hybrid aligned nematics strongly anchored on the planar side. *Mol. Cryst. Liq. Cryst.* **1990**, *179*, 425–433. [CrossRef]
13. Madhusudana, N.V.; Durand, G. Linear flexo-electro-optic effect in a hybrid aligned nematic liquid crystal cell. *J. Phys. Lett.* **1985**, *46*, 195–200. [CrossRef]
14. Strigazzi, A. On the critical thickness of a twisted nematic cell. *J. Phys.* **1985**, *46*, 1507–1512. [CrossRef]
15. Yamaguchi, R.; Sato, S. Continuous grey scale image printing on the liquid crystal cell. *Appl. Phys. Lett.* **2005**, *86*, 031913. [CrossRef]
16. Uchida, T.; Wada, M. Guest-Host Type Liquid Crystal Displays. *Mol. Cryst. Liq. Cryst.* **1981**, *63*, 19–44. [CrossRef]
17. Allia, P.; Oldano, C.; Trossi, L. Polarization transfer matrix for the transmission of light through liquid-crystal slabs. *J. Opt. Soc. Am. B* **1988**, *5*, 2452–2461. [CrossRef]
18. Lien, A. Extended Jones matrix representation for the twisted nematic liquid-crystal display at oblique incidence. *Appl. Phys. Lett.* **1990**, *57*, 2767. [CrossRef]

symmetry

MDPI

Article

Generation of Geometric Extra Phase and Accompanying Temporal Effects in Asymmetric Optically Compensated IPS-LCDs and FFS-LCDs

Shunsuke Kobayashi [1,*], Tomohiro Miyama [1], Hidenari Akiyama [2], Atsushi Ikemura [3] and Michio Kitamura [3]

1 Department of Electronic Engineering, Sanyo-Onoda City University, Yamaguchi 756-0884, Japan; miyama@rs.socu.ac.jp
2 DIC Corp., Saitama 362-8577, Japan; hidenari-akiyama@ma.dic.co.jp
3 Shintech, Yamaguchi 742-1512, Japan; ikemura@shintech.jp (A.I.); kitamura_ds@shintech.jp (M.K.)
* Correspondence: lcd.kobayashi@gmail.com

Abstract: This paper reports the generation of an extra phase and the accompanying temporal effects in an asymmetric optically compensated in-pane-switching (IPS) liquid crystal (LC) system and a fringe-field-switching (FFS) liquid crystal display (LCD) exhibiting a twofold faster response speed in the switching-off process compared with that in single and symmetric IPS-LCDs and FFS LCDs for the first time. To explain the experimental results, we derived an approximate analytical formula for the optical output intensity that includes an extra phase advancement and conducted simulations to achieve normally black operation using a dynamic optical retarder.

Keywords: IPS; FFS-LCD; asymmetric optical -LCD system; optical compensation; response time

Citation: Kobayashi, S.; Miyama, T.; Akiyama, H.; Ikemura, A.; Kitamura, M. Generation of Geometric Extra Phase and Accompanying Temporal Effects in Asymmetric Optically Compensated IPS-LCDs and FFS-LCDs. *Symmetry* **2021**, *13*, 1143. https://doi.org/10.3390/sym13071143

Academic Editor: Vladimir A. Stephanovich

Received: 10 April 2021
Accepted: 7 June 2021
Published: 26 June 2021

1. Introduction

Liquid crystal displays (LCDs) are widely used as information displays for televisions, computer monitors, and various instruments. Among the various LCD operation modes, in-plane-switching (IPS) [1] and fringe-field-switching (FFS) [2] are currently widely used because of the excellent wide viewing angles of IPS and FFS displays. However, their response speed needs further improvement. To solve this problem, several methods have been adopted (1) using liquid crystal (LC) materials with low viscosity [3,4], (2) devising electrode structures [5,6], (3) doping with chiral agents [7], (4) doping with nanoparticles [7], (5) using ferroelectric LC materials [8,9], (6) using an LCD with a narrow gap cell [10], and (7) overdriving to enhance the response speed in the switching-on process [11].

Our approach reported in the present paper represents an alternative to the above-mentioned approaches. In a preliminary previous work, we reported the electro-optical performances of asymmetric optically compensated tunable birefringent (TB) mode LCD exhibiting two-fold fast response speed in the switching off process [7]. For this reason, this present work was conducted aiming at the extension of the previous work to this proposed asymmetric optically compensated IPS and FFS LCDs (Asymmetric OC-IPS and FFS-LCD). The present work is related to the geometry phase, originally described by Pancharatnum [12] and Berry [13] in their historically representative papers. The present paper also updates previous work by introducing a dynamic optical retarder [7] for realizing electro-optical characteristics with by introducing a dynamic retarder, we realized normally black state normally black operation, accompanying with a simulation utilizing a simulator, SHINTECH LCD MASTER. In conjunction with a simulation using a SHINTECH LCD MASTER simulator, we introduce a dynamic retarder, to realize normally black operation.

2. Theory

2.1. Optical Configuration of our Asymmetric IPS-LC System

In Figure 1, angle φ_1 is the switching azimuthal angle of LC molecules and φ_2 is the setting azimuthal angle of a compensator (+A Plate) whose angle is set to $\varphi_2 = 3\pi/4 - \alpha$, which can be controlled using a dynamic retarder, when we use an IPS cell for the compensator.

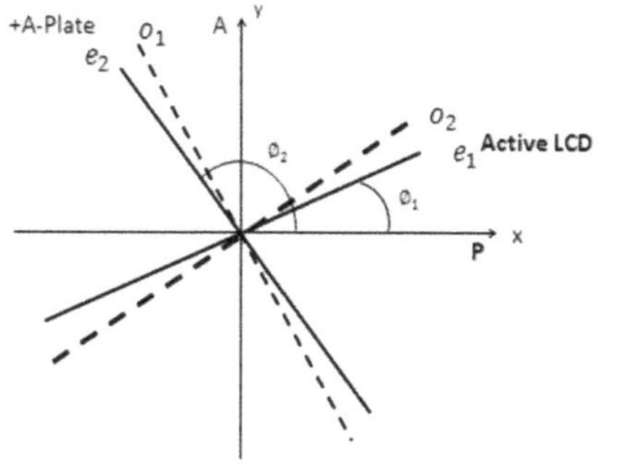

Figure 1. Optical configuration of our asymmetric IPS-LCD (FFS-LCD) system.

The whole system is sandwiched between crossed polarizers, where the incident light field into the LCD system is polarized in the x-direction and the transmitted wave is polarized in the y-direction. Labels e_1 and o_1 in Figure 1 indicate that the axes of the extra-ordinary wave and the ordinary wave, respectively, and labels e_2 and o_2 indicate those of the compensator, respectively.

When $\varphi_2 = 3\pi/4$ and $\varphi_1 = \pi/4$, the system is symmetric and generates a black state; under these conditions, the extra phase is not generated and the high response speed is not achieved.

2.2. Analytical Derivation of Normalized Optical Transmission of a Single IPS-Cell

Using the system shown in Figure 1, we here carry out a 2 × 2 Jones matrix calculation [14] for I_1 of a single cell.

The input electrical field is

$$\begin{bmatrix} 1 \\ 0 \end{bmatrix} E$$

We then have a projected field in the *x*-*y* coordinate:

$$\begin{bmatrix} E_1(x) \\ E_1(y) \end{bmatrix} = \begin{bmatrix} c_1 & -s_1 \\ s_1 & c_1 \end{bmatrix} \begin{bmatrix} -G_{11} & 0 \\ 0 & G_{12} \end{bmatrix} \begin{bmatrix} c_1 & s_1 \\ -s_1 & c_1 \end{bmatrix} \begin{bmatrix} 1 \\ 0 \end{bmatrix} E_0 \sin(\omega t) \tag{1}$$

where c_1 and s_1 are $\cos(\varphi_1)$ and $\sin(\varphi_1)$, respectively, here $G_{11} = \exp(-i\,\Gamma_{11})$ and $G_{12} = \exp(-i\,\Gamma_{12})$; in addition, $\delta_1 = \Gamma_{11} - \Gamma_{12}$, where each of which is the phase delay in the e_1 and o_1, from the incident wave, respectively.

We have the *x*-*y* components such that:

$$E_1(x) = [(c_1)^2 \exp(-i\delta_1) + (s_1)^2]E_0, \tag{2}$$

$$E_1(y) = [s_1 c_1(\exp(-i\delta_1) - 1)]E_0, \tag{3}$$

Then,

$$E_1(y) = E_0 \sin(\varphi_1)\cos(\varphi_1) \exp(-i\delta_1). \tag{4}$$

$$= 2E_0 \sin(2\varphi_1)\exp(-i\delta_1). \tag{5}$$

Thus, we have the optical output intensities and the normalized optical transmission for the single cell:

$$I_1 = I_0 \sin^2(2\varphi_1)\sin^2(\delta_1/2). \tag{6}$$

The normalized transmission is then

$$T_1 = \sin^2(2\varphi_1). \tag{7}$$

A $\delta_1 = 1.3\pi$ for n-IPS and n, p-FFS and 1.04π for p-IPS were determined based on the properties of the LC materials and the LCD cell fabrication details (Section 3.3).

2.3. Derivation of Normalized Optical Transmission Intensity of OC-IPS with 2 × 2 Jones Matrix Calculations

We here derive the equation for the proposed asymmetric OC-IPS and FFS shown in Figure 1. This derivation is carried out by replacing the c_1 and s_1 terms in Equation (1) with c_2 and s_2 and $\Gamma_{21} - \Gamma_{22}$. In addition, by inserting $E_1(x)$ and $E_1(y)$, we obtain

$$E_1(x) = (s_1)^2 - (c_1)^2 \text{ and } E_1(y) = -2s_1c_1$$

We derive $E_2(y)$ as follows:

$$\begin{bmatrix} E_2(x) \\ E_2(y) \end{bmatrix} = \begin{bmatrix} c_2 & -s_2 \\ s_2 & c_2 \end{bmatrix}\begin{bmatrix} -G_{21} & 0 \\ 0 & G_{22} \end{bmatrix}\begin{bmatrix} c_2 & s_2 \\ -s_2 & c_2 \end{bmatrix}\begin{bmatrix} E_1(x) \\ E_1(y) \end{bmatrix} \tag{8}$$

where δ_2 is again the difference in phase delay between the e_2 and o_2 axes:

$$\delta_2 = \Gamma_{21} - \Gamma_{22}.$$

We then have:

$$E_2(y) = 2[s_2c_2(c_1{}^2 - s_1{}^2) - s_1c_1(c_2{}^2 - s_2{}^2)]\exp(-i\delta_2). \tag{9}$$

And $\delta_2 = \delta_1 = 1.3\pi$ for n-IPS and n, p-FFS and 1.04π for p-IPS as evaluated based on the physical properties and the LCD cell fabrication conditions described in Section 3.3. Then, with $\sin^2(\omega t) = 1/2$ and without $\exp(-i\delta_2)$ to normalize the optical intensity and transmission.

$$I_2 = I_0 \sin^2(2\varphi_2 - 2\varphi_1)\sin^2(\delta_1/2)\sin^2(\delta_2/2) \tag{10}$$

$$T_2 = \sin^2(2\varphi_2 - 2\varphi_1), \tag{11}$$

If we substitute $\varphi_2 = \pi/4 - \alpha$ into Equation (11), we obtain an equation with a phase-advancement shift, of 2α:

$$T_2 = I_0 \cos^2(2\varphi_1 + 2\alpha) \tag{12}$$

In our system, the symmetry breaking introduces a phase shift, which leads to a fast response in the switching-off process. This effect is shown later in Figure 3. When LC-cell and the A-plate are crossed, $\varphi_1 = \varphi_2 = \pi/4$ and $\delta_1 = \delta_2 = 1.3\pi$, the symmetric system then produces the black (dark) state and no ultrafast switching process occurs.

3. Results and Discussion

3.1. Optical Configurations and Switching of Single and OC-IPS, and OC-FFS Cells with $\Delta\varepsilon > 0$ and $\Delta\varepsilon < 0$

Figure 2a,b show the optical configurations and switching of single and OC-IPS and OC-FFS-LCD cells used in the present research, where p and n denote $\Delta\varepsilon > 0$ and $\Delta\varepsilon < 0$, respectively.

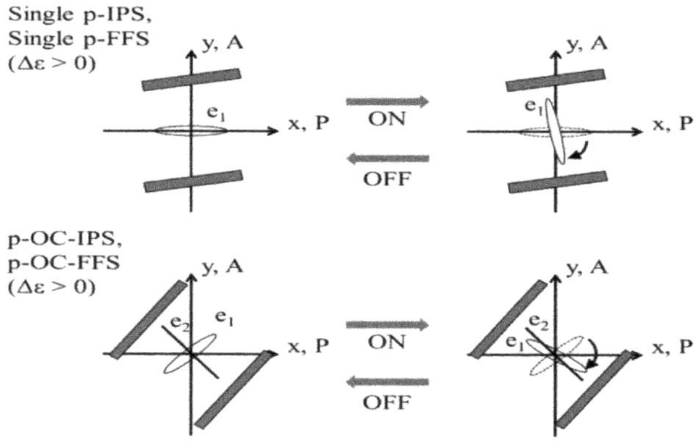

(**a**) Switching of p-IPS and p-FFS

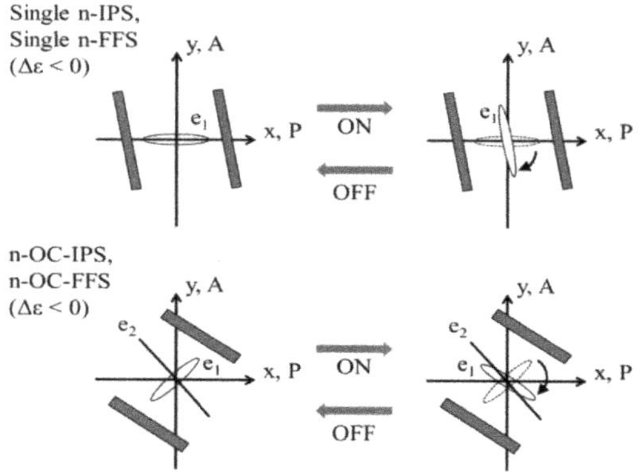

(**b**) Switching of n-IPS and n-FFS

Figure 2. Optical configuration of switching in single and optically compensated IPS- and FFS-LCDs with negative dielectric anisotropy.

We next explain the switching process in our OC-LCDs, along with the contents of Figure 2.

Figure 3 shows simulation results for the normalized transmission and compares T_1 (I_1) with T_2 (I_2) against the switching angles calculated using Equations (7) and (12). The top and central parts in this figure show the behaviors in the switching-off process, and the bottom and right parts show the behaviors in the switching-on process. Thus, the φ_1 for T_2 (I_2) starts from a degree of $-\alpha$, and the φ_1 for the T_1 (I_1) starts from $-45°$.

The results in Figure 3 are interpreted as follows (1) The T_2 (I_2) has a phase advancement of α over the $I_1(T_1)$, (2) According to the top area, the T_2 (I_2) starts the decay process with a finite inclination, whereas the T_1 (I_1) starts the decay process with no inclination. This decay in the absence of inclination is attributed to the fast response in the decay process, which was experimentally demonstrated in the present research. This phenomenon is a purely optical effect, where the decay process occurs as the common molecular relaxation for both the symmetric and asymmetric systems. (3) The I_2 (T_2) has a finite value at the right bottom because of energy conservation, which means that the operation is not normally black. The realization of a normally black operation will be described in Section 3.4.

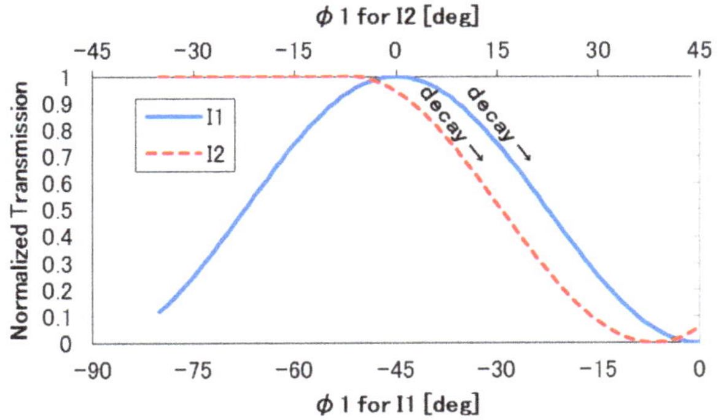

Figure 3. Simulation results of a comparison of switching characteristics of I_1 (T_1) with I_2 (T_2) starting from the switching off process.

3.2. V-T Curves of Single p-IPS and p-OC-IPS Devices

Figure 4 shows the experimentally obtained transmittance-voltage (V-T)-curves of IPS LCDs, as recorded at 25 °C using a DMS-703 (Autronic Melchers, GMb) instrument. Interestingly, in the symmetric compensation system, the introduction of optical compensation introduces a wide operating voltage range, as shown in Figure 4. Regarding the V-T curve, refer to [14], where the Freedericksz transition is described.

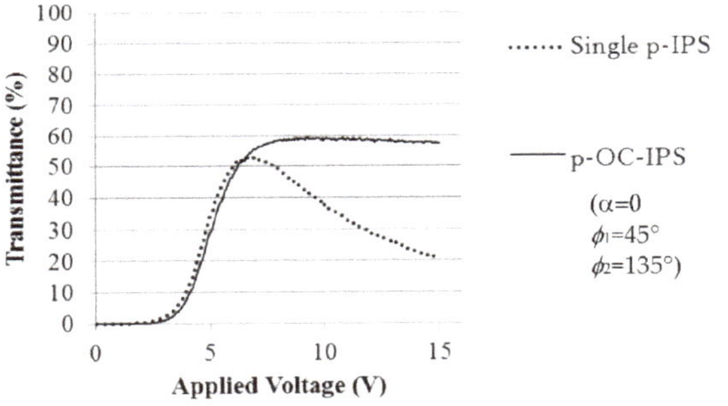

Figure 4. Experimentally obtained V-T curves for p-IPS ($\Delta\varepsilon > 0$) and p-OC-IPS devices with $\alpha = 0$ (symmetric).

Figure 5 shows the V-T curves of the p-OC-IPS device without optical compensation ($\alpha = 0$) and the p-OC-IPS device with optical compensation, ($\alpha = -6.8°$). The proposed asymmetric OC-IP produces $T = 0$ at 4 V; which means a non-normally black operation. In general, an LCD is operated normally black or normally white; thus, a dynamic optical compensator, which has been updated from the device described in our previous work [7], was used (see Section 3.5).

3.3. Experimental Results of Response Time

The physical properties of the LC materials and the specifications of the LCD cells used in the present work are described as follows: The cell gap of the sample cells was 3.5 μm and the wavelength of the optical light source was 589 nm. The anisotropy of refractive index,

Δn (at 589 nm, 25 °C) and $\Delta\varepsilon$ (25 °C) (After Datasheet of DIC)

p-IPS	0.088	6.2
n-IPS	0.110	−4.1
p-FFS	0.110	11.9
n-FFS	0.110	−4.1

Using these quantities,

The experimental results in Tables 1 and 2, indicate that the response time of the proposed asymmetric OC-IPS and OC-FFS devices are reduced by 42–56% compared with those of single cells and symmetric OC-IPS and OC-FFS devices. Approximately the same results were obtained for the OC-FFS device, and we will report the corresponding data elsewhere. These results indicate that the response time in the switching-on process will be greatly reduced by adopting the overdriving technique [12].

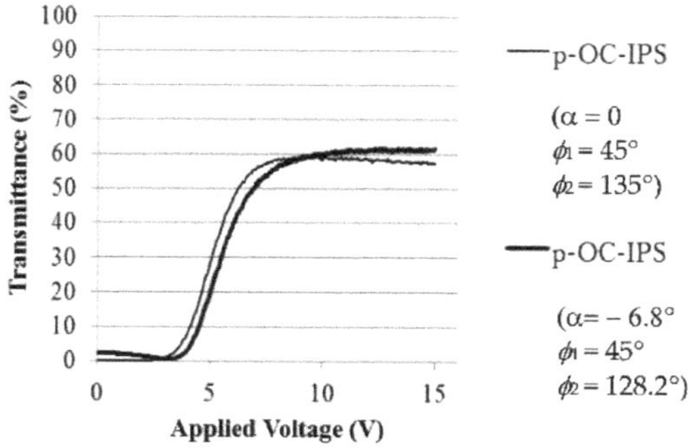

Figure 5. Experimentally obtained *V-T* curves of a p-OC-IPS device without optical compensation, ($\alpha = 0$) and a p-OC-PS device with optical compensation, ($\alpha = -6.8°$).

Table 1. Results of response time for the n-OC-IPS device.

	Single n-IPS	n-OC-IPS ($\alpha = 0°$)	n-OC-IPS ($\alpha = -6.8°$)
τ_{off} (msec)	25.6	25.1	14.9 (42% decrease)
V_{op} (V)	6.8	9.1	15.0

Table 2. Results of response time for the p-OC-IPS device.

	Single p-IPS	p-OC-IPS ($\alpha = 0°$)	p-OC-IPS ($\alpha = -6.8°$)
τ_{off} (msec)	18.3	14.1	8.1 (56% decrease)
V_{op} (V)	4.0	4.5	5.5

3.4. Temporal Derivatives of I_1 and I_2 and Their Comparison

We here derive temporal derivatives of I_1 and I_2

$$\frac{\partial I_1}{\partial t} = \frac{\partial I_1}{\partial \varphi_1}\frac{\partial \varphi_1}{\partial t} = 2I_0 \sin(4\varphi_1)\frac{\partial \varphi_1}{\partial t} \tag{13}$$

For an OC-IPS:

$$\begin{aligned} \tfrac{\partial I_2}{\partial t} &= \tfrac{\partial I_2}{\partial \varphi_1}\tfrac{\partial \varphi_1}{\partial t} = -2I_0 \sin(4\varphi_2 - 4\varphi_1)\tfrac{\partial \varphi_1}{\partial t} \\ &= -2I_0 \sin(4\varphi_1)[\cos(4\alpha) - \sin(4\alpha)\cot(4\varphi_1)]\tfrac{\partial \varphi_1}{\partial t} \end{aligned} \tag{14}$$

Thus, we obtain a formula, the F-function, by taking the ratio between Equations (13) and (14) as follows, where the two minus signs cancel each other:

$$F = \cos(4\alpha) - \sin(4\alpha)\cot(4\varphi_1)$$

The F-function ($\frac{\partial I_2}{\partial t} / \frac{\partial I_1}{\partial t}$) is shown in Figure 6.

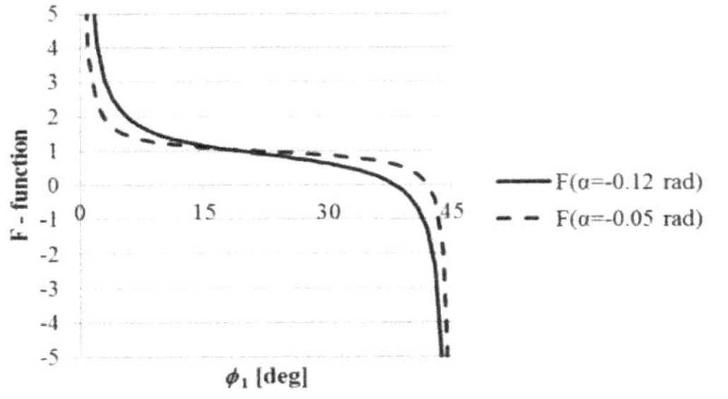

Figure 6. The F-functions with $\alpha = -0.12$ rad and $\alpha = -0.05$ rad.

Figure 6 shows that the temporal derivative of I_2 is much larger than that of I_1 during the switching-off process, consistent with the experimental results obtained in the present work. However, it also shows that this effect terminates at a switching angle of 15° and that the relation is thereafter inverted. However, if necessary, a dynamic compensator can be used to eliminate the phase-advancement after the decay process for the switching-on process, the removal of the applied voltage causes an overvoltage, resolving this problem [11].

3.5. *Dynamic Compensator*

A dynamic compensator changes the azimuthal angle of the e-axis, $\varphi_2 = 3\pi/4 - \alpha$, by applying a computer-controlled electric voltage, V_{appl}, according to the following equation:

$$\alpha = \alpha_0 \exp\,[1 - kV_{\text{th}}/V_{\text{appl}}],\ k > 1\ (k = 2.5) \tag{15}$$

(1) For the switching-on operation (starting from $V_{\text{appl}} \approx 0$), normally black operation will be induced with $\alpha = 0$ ($\varphi_2 = 3\pi/4$).
(2) For the switching-off operation, where $V_{\text{appl}} = kV_{\text{th}}$ ($k = 2.5$), then $\alpha = \alpha_0$ (~7°), extra phase, α, is generated.

This practice is demonstrated in Figure 7.

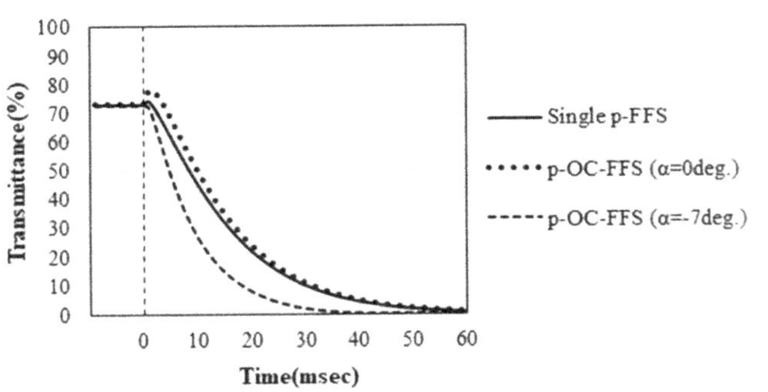

Figure 7. Demonstration of normally black mode operation using a dynamic compensator.

4. Conclusions

1. The optical output intensities were formulated using a slightly modified Jones matrix calculation, and the results showed that an extra phase accompanying temporal behavior is generated in the switching angle for the proposed asymmetric optically compensated IPS -LCD and FFS-LCD systems.
2. Experiments showed that the optically compensated IPS-LCD has a broad dynamic range beyond the sine-squared oscillation in the ordinary IPS-LCD.
3. Simulation results revealed the existence of phase advancement in the optical output intensity of the proposed asymmetric OC-IPS relative to that of symmetric OC-IPS, depending on the selected optical retardation in the compensator.
4. Experimentally measured values of the response time of the proposed asymmetric OC-IPS-LCD and OC-FFS were reduced by 42–56% compared with those of symmetric OC-IPS, and OC-FFS, LCDs, and single IPS and FFS-LCDs.
5. Temporal derivatives of the optical intensity for the proposed asymmetric OC-IPS were shown to be larger than those of the symmetric OC-IPS.
6. By adopting a dynamic compensator, we demonstrated normally black operation for the proposed asymmetric OC-IPS.

Author Contributions: Conceptualization, S.K.; derivatization of analytical formulas of optical transmission, S.K.; writing, original draft preparation, S.K. and T.M.; sample fabrication and experimental measurements, H.A. and T.M.; simulations, M.K., A.I., and T.M.; All authors have read and agreed to the published version of the manuscript.

Funding: This work was partially supported by a grant from the Ministry of Education, Culture, Sports, Science and Technology, Japan (MEXT) as a project of Yamaguchi Green Materials Cluster.

Institutional Review Board Statement: Not applicable.

Informed Consent Statement: Not applicable.

Data Availability Statement: Japanese patent 6717323 (2020.7.1). Title: Method for improving response speed and liquid crystal display device, International release No.PCT/JP2016/086379. Invented by Shunsuke Kobayashi, Hidenari Akiyama, and Kenzo Takeishi.

Acknowledgments: The authors will wish to thank Haruyoshi Takatsu and Hiroshi Hasebe at DIC Corp. for their kind and enthusiastic collaboration in LC materials research. And the authors are indebted to Shoichi Ishihara at OIT for executing the preparation of this paper.

Conflicts of Interest: The authors declare no conflict of interest.

References

1. Oh-e, M.; Kondo, K. Electro-optical characteristics and switching behavior of the in-plane switching mode. *Appl. Phys. Lett.* **1995**, *67*, 3895–3897. [CrossRef]
2. Lee, S.-H.; Lee, S.-L.; Kim, H.-Y. Electro-optic Characteristics and switching principle of a nematic liquid crystal cell controlled by fringe-field-switching. *Appl. Phys. Lett.* **1998**, *73*, 2881–2883. [CrossRef]
3. Bremer, M.; Klassen-Memmer, M.; Tarumi, K. Development of Liquid Crystal Displays and Its Materials. In *Honor of Shunsuke Kobayashi's 80 Years Birth Day*; Kwok, H.S., Naemura, S., Ong, L.H., Eds.; World Scientific: Singapore, 2013; Chapter 4, pp. 97–131.
4. Takatsu, H. Liquid Crystal Materials. In *High Quality Liquid Crystal Displays and Smart Devices*; Ishihara, S., Kobayashi, S., Ukai, Y., Eds.; IET: London, UK, 2018; Chapter 18, pp. 315–329.
5. Park, Y.-J.; Kim, Y.-W.; Choi, T.-H.; Yoon, T.-H. Sub-millisecond switching of a nematic liquid crystal cell aided by a vertical bias field. In Proceedings of the IDW 2014, Niigata, Japan, 3–5 December 2014; Volume 1, pp. 86–87.
6. Matsushima, T.; Yang, Y.B.; Takizawa, K. New Fast Response Time In-Plane Switching Liquid Crystal Mode. In *Digest of Technical Papers*; No.43.2; SID: San Jose, CA, USA, 2015; pp. 648–651.
7. Kobayashi, S. Progress on Liquid Crystal Science and Technology. In *Honor of Shunsuke Kobayashi's 80th Birthday*; Kwok, H.S., Naemura, S., Ong, L.H., Eds.; World Scientific: Singapore, 2013; Chapter 1.
8. Clark, N.A.; Lagerwall, S. Sub-microsecond bistable electro-optic switching in liquid-crystals. *Appl. Phys. Lett.* **1980**, *36*, L1517–L1521. [CrossRef]
9. Ishihara, S.; Kobayashi, S.; Ukai, Y. Chapter 16 Sequential Full Colour Displays Using Polymer-Stabilized Ferroelectric LCD. In *High Quality Liquid Crystal Displays and Smart Devices*; IET: London, UK, 2018; pp. 315–329.

10. Kobayashi, S.; Kundu, B.; Miyama, T.; Shiraishi, Y.; Sawai, H.; Toshima, N.; Okita, M.; Takeuchi, K.; Takatsu, H.J. Narrow-gap field-sequential TN-LCD with and without nanoparticle doping. *Inf. Disp.* **2011**, *19*, 19787–19792. [CrossRef]
11. Okumura, H.; Baba, M.; Kobayashi, H. Level-Adaptive Overdrive (LAO) Technology. In Proceedings of the IDW2002, Hiroshima, Japan, 4–6 December 2002; AMD1/FMC2-2. pp. 207–210.
12. Pancharatnam, S. Generalized Theory of Iterrference, and Its Applications Part 1 Coherence pencils. *Proc. Indian Acad. Sci.* **1956**, *44*, 247–262. [CrossRef]
13. Berry, M.V. Quantal phase factors accompanying adiabatic changes. *Proc. R. Soc. Lond. A* **1984**, *392*, 45–47.
14. Yang, D.K.; Wu, S.T. *Fundamentals of Liquid Crystal Devices*; Wiley SID; Anthony Rowe Ltd.: Chippenham, Wiltshire, UK, 2006; p. 74.

symmetry

MDPI

Article

A Novel Technique for Determination of Residual Direct-Current Voltage of Liquid Crystal Cells with Vertical and In-Plane Electric Fields

Msanobu Mizusaki [1,*] and **Shoichi Ishihara** [2]

1 Development Group, Sharp Display Technology Corporation, Tenri 632-8567, Japan
2 Department of Electrical and Electronic Engineering, Graduate School of Engineering, Osaka Institute of Technology, Osaka 535-8585, Japan; shoichi.ishihara@oit.ac.jp
* Correspondence: mizusaki.masanobu@sharp.co.jp; Tel.: +81-50-5433-4691

Abstract: Generation of residual direct-current (DC) voltage (V_{rDC}) induces serious image sticking of liquid crystal displays (LCDs). In this study, a novel technique to determine the V_{rDC} of LC cells is proposed. We found that the V_{rDC} could be determined from a current-voltage (*I-V*) curve obtained by the application of triangular voltage. In the case of a vertically aligned twisted nematic (VTN) mode LC cell, where a vertical electric field is applied, the *I-V* curve shows maximum and minimum current peaks owing to rotation of an LC director, and the V_{rDC} is able to be determined from an average value of the two peaks. On the other hand, in the case of a fringe field switching (FFS) mode LC cell, where an in-plane (lateral) electric field is applied from comb electrodes, the current peaks derived from the rotation of the LC director do not appear. Therefore, we could not adopt the same way with that of the VTN mode LC cell. However, we found that there were two minimum current peaks derived from minimum capacitances of the FFS mode LC cell, and could determine the V_{rDC} by using these two current peaks. The proposed technique would be useful for the evaluation of the V_{rDC} of the LCDs, where the electric field is applied both vertically and laterally.

Keywords: liquid crystal display; residual direct-current voltage; *I-V* curve; triangular voltage application; vertically aligned twisted nematic (VTN) mode; fringe field switching (FFS) mode

Citation: Mizusaki, M.; Ishihara, S. A Novel Technique for Determination of Residual Direct-Current Voltage of Liquid Crystal Cells with Vertical and In-Plane Electric Fields. *Symmetry* **2021**, *13*, 816. https://doi.org/10.3390/sym13050816

Academic Editor: Raffaele Barretta

Received: 5 April 2021
Accepted: 29 April 2021
Published: 7 May 2021

1. Introduction

In recent years, the progress for the display performance of liquid crystal displays (LCDs) is remarkable with high contrast ratio, wide viewing angle, and fast response owing to significant advancement of liquid crystal (LC) alignment control technology [1–5]. Therefore, the application of the LCDs extends to many fields: smartphone, watch, tablet, television, automobile display, and so on. With the significant progress of the above image quality, there is a possibility that quite a small level of image defect becomes observable [6–10]. Image sticking is a particularly serious problem because the human eye generally distinguishes quite a small level of image sticking. Hence, a solution to this problem is necessary. It is known that image sticking is the phenomenon in which the previous pattern is visible whilst the next pattern is addressing [11]. One of the reasons for the image sticking of the LCDs is the generation of residual direct-current voltage (V_{rDC}) [12–18]. The V_{rDC} is the internal voltage being generated in the LC cell and is estimated to be derived from the following mechanisms [14–18]. One of the mechanisms is that an impurity ion in an LC layer of an LC cell adsorbs on a surface of an alignment layer under application of external DC offset voltage, and then the adsorbed ion does not desorb from the surface for a long period even after the DC offset voltage is removed [14–17]. Hence, the adsorbed ion on the surface of the alignment layer is one of the roots of the V_{rDC}. The other mechanism of the V_{rDC} is the generation of charged voltage in the LC cell, as depicted in Figure 1 [18]. The LC cell generates the charged voltage by the application of the DC voltage because both the LC layer and the alignment layer are composed of dielectric compounds. During the open circuit period after a period of discharge, the charged voltages generated both in the LC layer and the alignment layer relax. However, the relaxation time for the charged voltage of the alignment layer is generally long because the resistance of the alignment layer is relatively high. Thus, the residual charged voltage in the LC cell is another root of the V_{rDC} [18].

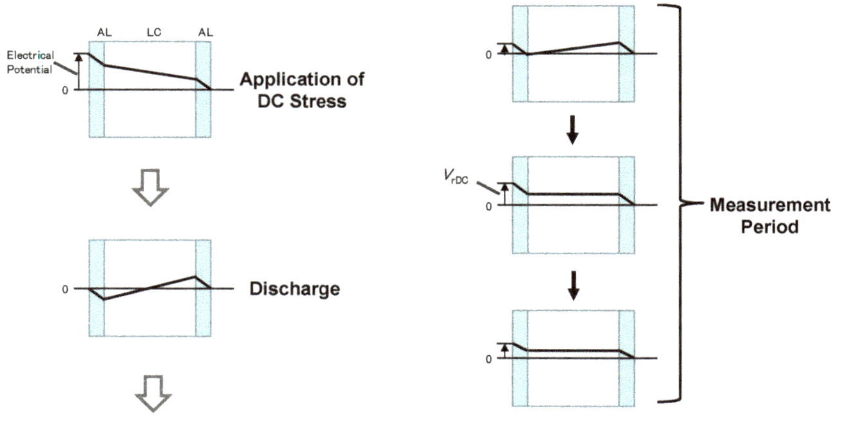

Figure 1. Shift of the charged voltage of the LC cell during the periods of the DC voltage application, discharge, and open circuit [18].

So far, several measurement techniques have been proposed for the evaluation of V_{rDC}, as listed in Table 1. One is the capacitance-voltage (*C-V*) hysteresis technique: the V_{rDC} is determined by the *C-V* curve shift after applying the DC voltage [19,20]. In this technique, square wave voltage is applied to the LC cells [19]. The second technique is the flicker minimization technique: the V_{rDC} is determined by minimizing an optical flicker by optimizing the DC offset voltage [16,21]. The last one is the dielectric absorption technique: the V_{rDC} is determined by the residual charged voltage after application of the DC voltage and its discharged process [18]. In these techniques, we usually used the flicker minimization technique for evaluation of the V_{rDC} of the LC cells such as twisted nematic (TN), electrically controlled birefringence (ECB), vertically aligned (VA), and VA-TN (VTN) mode cells because this optical evaluation technique was empirically coincidence with the image sticking of the LCDs. Figure 2a shows an example of the flicker generation due to the V_{rDC}. Though the LC cell is applied AC voltage without the DC offset, the existence of the V_{rDC} induces frequently different voltage applications to the LC layer. Therefore, the transmittance changes frequently, inducing the flicker. However, the flicker is not detected when the V_{rDC} is 0 V, as shown in Figure 2b. Regarding this technique, there is a problem that it is not easy to determine the accurate V_{rDC} because the flicker is gradually shifted during the adjustment of the flicker minimum point. In particular, it is extremely difficult to determine the V_{rDC} by this technique for the LC cell with the lateral electric field such as in-plane switching (IPS) or fringe-field switching (FFS) modes because the generation of the flicker is derived not only from the V_{rDC} but also from a flexoelectric field [22–27]. In this study, we will focus on the V_{rDC}, and the purpose is to propose an alternate technique to determine the V_{rDC} of the LC cell especially with the lateral electric field more accurately. In particular, we will focus on the FFS mode because most of the LCDs adopted in various applications are the FFS mode. The proposed technique is the current-voltage (*I-V*) curve shift technique by the application of a triangular voltage to the LC cells. By applying the triangular voltage to the LC cell, the current shift can be monitored owing to capacitance shift, resistance originated from Ohm's law, and motion of the LC directors [18]. Therefore, the *I-V* curve obtained by the application of the triangular voltage gives us various properties of the LC cell.

Table 1. Conventional evaluation techniques for V_{rDC}.

	C-V Hysteresis [19,20]	Flicker Minimization [16,21]	Dielectric Absorption [18] (Refer to Figure 1)
Measuring method	Electrical	Optical	Electrical
Measurement item	Shift of C-V curve	Offset voltage to minimize optical flicker	Change of electrical potential
Correlation with the image sticking	High	High	Middle
Convenience for measurement	Easy	Difficult	Easy
Adaptivity to the LC cells with vertical electric field (TN, ECB, VA, VTN modes)	Applicable	Applicable	Applicable
Adaptivity to the LC cells with in-plane electric field (IPS and FFS modes)	No Data	Difficult	Applicable

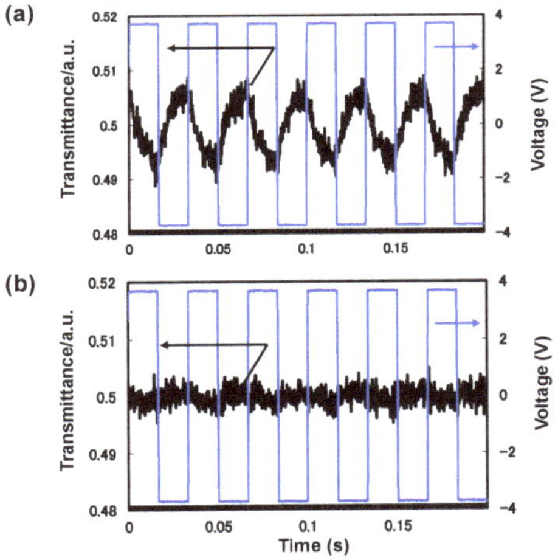

Figure 2. Schematic illustrations for (**a**) the generation of the flicker due to generation of V_{rDC}, and (**b**) the minimization of the flicker.

2. Two Types of the LC Cells

We selected two types of LC cells; one is the VTN mode cell where the electric field is applied vertically, and the other is the FFS mode cell where the electric field is applied laterally on one substrate with electrodes. These are shown in Figure 3a,b, respectively.

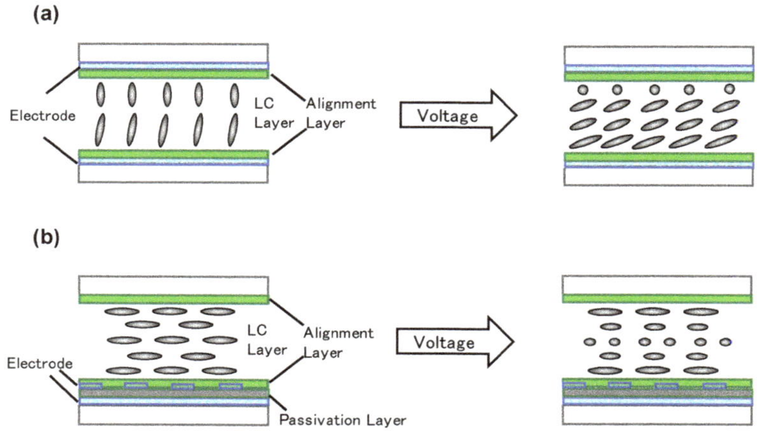

Figure 3. (**a**) The LC cell with vertical electric field owing to the upper/lower electrodes, and (**b**) the LC cell with the in-plane electric field owing to the comb electrodes on one side of the pair of the substrates.

In the case of the VTN mode LC cell, LC directors inside the LC layer rotate vertically and uniformly due to the application of voltage (Figure 3a). On the other hand, the FFS mode LC cell shows the lateral rotation of the LC directors only in the neighborhood area of the electrodes under application of the voltage (Figure 3b) [5,28]. The different behaviors of the LC directors in two LC modes are explained in the following. In the case of the VTN mode, the electric field appears uniformly and symmetrically in the LC layer as the simulation result is shown in Figure 4a,b. The physical properties of the LC material used for this simulation are listed in Table 2, and the layer thickness and pretilt angle of the LC materials are 4.0 µm and 88.5°, respectively. In contrast, the electric field appears non-uniformly for the FFS mode, and then the distribution of the LC director under application of the voltage is non-uniform in the in-plane area, as shown in Figure 5a,b. The simulation was performed under the line/space = 3.5 µm/4.5 µm, the dielectric constant of the passivation layer is 7, and its thickness is 200 nm. In this simulation, the same LC material as that of the VTN mode is used, and the pretilt angle is 0.1°. The density of the electric field in the neighborhood area of the electrodes is high, and that of the electric field around the opposite substrate is low. In other words, the electric field of the LC cell with the comb electrodes shows non-uniform and asymmetry.

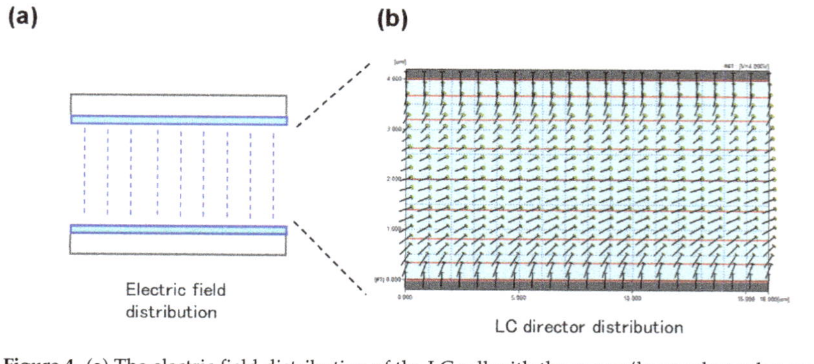

Figure 4. (**a**) The electric field distribution of the LC cell with the upper/lower electrodes, and (**b**) the LC director distribution of the VTN mode.

Table 2. Physical properties of the nega-LC material used in this study.

	Nega-LC
T_{NI} (°C)	90
$\varepsilon_{\parallel}$	3.6
$\varepsilon_{\perp}$	7.8
$\Delta\varepsilon$	−4.2
Δn	0.083
ρ (Ωcm)	1.8×10^{13}

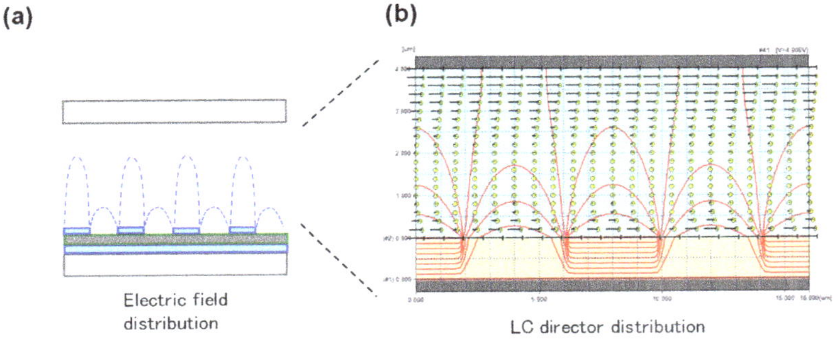

Figure 5. (**a**) The electric field distribution of the LC cell with the comb electrodes, and (**b**) the LC director distribution of the FFS mode. (Red lines mean equipotential lines.).

Figure 6 indicates the simulation result for rates of change of dC/dV as a function of the applied voltage in the VTN and FFS modes. The VTN mode shows a sharp and large peak at around 2.5 V, whereas the FFS mode shows only a small peak at around 1 V. This indicates that, in the VTN mode, the sharp increase for the capacitance of the LC cell appears due to response of the LC director under application of the voltage. On the other hand, in the FFS mode, only the slight increase for the capacitance of the LC cell appears probably due to the non-uniform electric field.

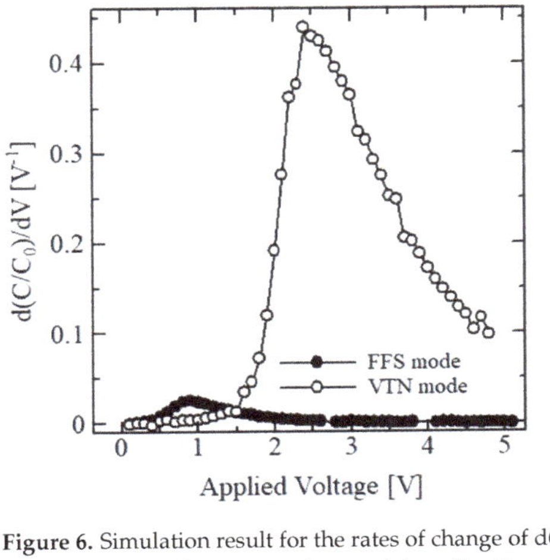

Figure 6. Simulation result for the rates of change of dC/dV in VTN and FFS modes as a function of applied voltage (C_0 is a capacitance of the cell with no applied voltage).

3. Preparation of LC Cells

The fabrication procedure of the VTN-LC cell is in the following. First, the polyimide material for the VA mode, purchased from JSR Corporation, was coated on a pair of glass substrates carrying transparent indium-tin-oxide (ITO) electrodes on the whole area. Then, linearly-polarized UV of 20 mJ/cm^2 was irradiated at a 30 °C atmosphere from an oblique direction for obtaining a slightly inclined pretilt angle [29,30]. Two glass substrates were assembled together for showing $\pi/2$ twist configuration of the LC directors with seal adhesive. The cell gap was maintained at 4.0 μm with glass fiber, and the LC material with negative dielectric anisotropy (nega-LC; Table 2) was injected into the empty cell by vacuum filling method, being obtained the VTN-LC cell.

In the case of the FFS-LC cell, horizontal alignment material purchased from JNC Corporation was coated on the pair of the substrates for the FFS mode. One side of the substrates carries the comb ITO electrode and a passivation layer, and the other substrate does not carry any electrode. The line and space of the electrodes were 3.5 μm and 4.5 μm, respectively. The thickness and dielectric constant of the passivation layer were 200 nm and 7.5, respectively. The linearly-polarized UV of 5 J/cm^2 was then irradiated from the normal direction to the substrates. Two glass substrates were assembled together with the seal adhesive. The cell gap was maintained at 4.0 μm, and the nega-LC material, whose physical properties are shown in Table 2, was injected into the empty cell by the same method described above. Finally, we obtained the FFS-LC cell. The parameters of the LC cells are listed in Table 3. Specific capacitance ("C") and resistance ("R") of the LC cells were determined by the current-voltage (*I-V*) curve by the application of 30 Hz triangular voltage from −10 V to +10 V at 25 °C. Voltage holding ratios (VHRs) were determined during an open-circuit of 16.61 ms after application of ±5 V for 60 μs, at 70 °C, with a 6254 type VHR meter developed by Toyo Corporation [18]. The specific capacitance of the VTN-LC cell is five times larger than that of the FFS-LC cell. On the other hand, the specific resistance of the VTN-LC cell is between 1/4 and 1/5 compared with that of the FFS-LC cell. This is probably due to the difference in the average distance of the electrodes between the VTN-LC and FFS-LC cells. The VHRs for both LC cells were almost equal, assuming that the amount of impurity ion in the LC layers of both LC cells are almost equal, as presented in Table 3.

Table 3. Parameters of the LC cells (VTN-LC cell *[1] and FFS-LC cell *[1]) used in this study.

	VTN-LC Cell	FFS-LC Cell
d (μm)	4.0	4.0
C *[2] (nF) *[2] /25 °C	1.3	0.28
R (GΩ) *[3] /25 °C	0.76	3.68
VHR/70 °C	98.4	98.1

*[1] Electrode area is 1 cm^2. *[2] Capacitance "C" is determined by the average capacitance value when +10 V is applied and −10 V is applied. *[3] Resistance "R" is determined from the average slopes of positive and negative sides when −10 V and + 10 V are applied.

4. Measurement of VA-LC Cell

4.1. I-V Curve

Figure 7 indicates the *I-V* curve of the VTN-LC cell obtained by the application of the 30 Hz triangular voltage from −10 V to +10 V at 25 °C. The slope of the *I-V* curve indicates an inverse of the resistance between the electrodes [18]. Generally, the slope of the *I-V* curve becomes small with increasing the resistance of the LC cell. The distance between the positive and negative current sides is attributed to the capacitance of the LC cell [18]. This increases with increasing the absolute value of applied voltage because the LC material shows the anisotropy of the dielectric constant [18]. Moreover, one can observe the maximum current peak in the range from +4 V to +7 V, and the minimum current peak in the range from −7 V to −4 V. These two peaks are attributed to a rotation of the LC director due to the anisotropy of the dielectric constant [18].

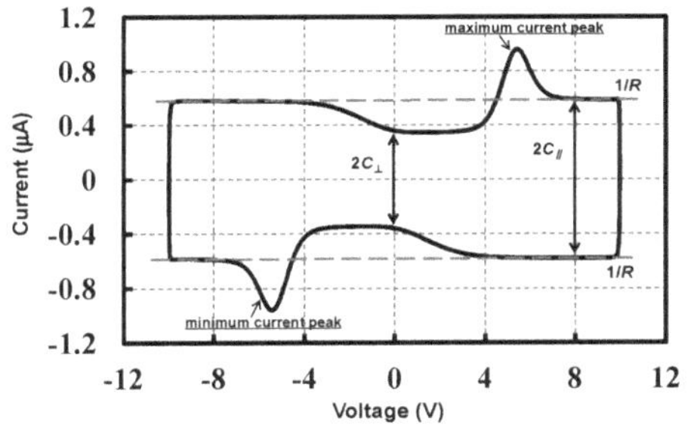

Figure 7. *I-V* curve for the VTN-LC cell and illustration of capacitance s ($C_\perp$ and $C_\parallel$), resistance (R), and maximum and minimum current peaks; the triangular voltage from −10 V to +10 V was applied with 30 Hz at 25 °C.

4.2. Determination of V_{rDC}

For the determination of the V_{rDC} from the *I-V* curve, the maximum and minimum current peaks derived from the rotation of the LC director are used. Before the application of DC offset voltage for stress, we can estimate that the V_{rDC} does not generate (V_{rDC} = 0 V). Figure 8a shows the *I-V* curve of the VTN-LC cell before application of the DC offset voltage, and we can observe the maximum and minimum current peaks at +5.38 V and −5.38 V, respectively. By using these two current peaks, the V_{rDC} can be determined from the average value, which is 0.00 V. In the next step, the DC offset voltage of +5 V for 180 min (3 h) was applied to the cell for the stress, and then the *I-V* curve was measured, which is shown in Figure 8b. The voltages of the maximum and minimum current peaks are +5.69 V and −5.25 V, respectively. The result that two current peaks are shifted indicates that the V_{rDC} is generated by the application of the DC offset voltage. We determined that the V_{rDC} was 0.22 V from the average voltage of the two current peaks. The determination of the V_{rDC} by using this method would be reasonable because the rotation of the LC director is induced by the application of the voltage to the LC layer. In the case that the internal DC voltage is generated in the LC layer, the application voltage to the cell would be shifted. We decide to call the method the "*I-V* curve shift" technique.

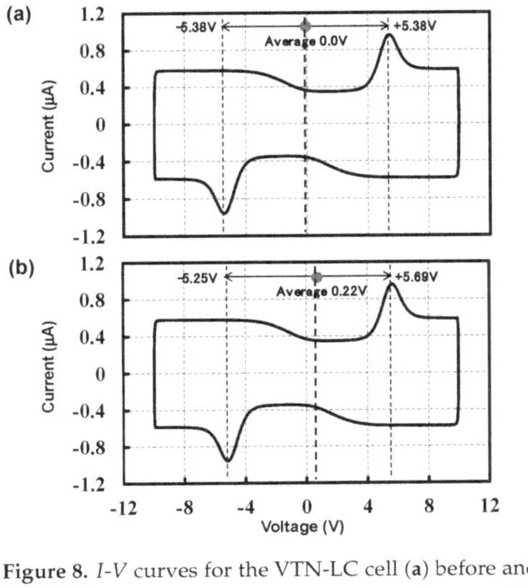

Figure 8. *I-V* curves for the VTN-LC cell (**a**) before and (**b**) after application of the +5 V DC offset voltage for 3 h (180 min) as the stress test; (condition of the measurement) the triangular voltage from −10 V to +10 V was applied with 30 Hz at 25 °C.

4.3. Comparison of V_{rDC} Determined by I-V Curve and Flicker Minimization

The comparison of the V_{rDC} determined from the *I-V* curve shift technique proposed in this report and the flicker minimization was carried out under application of the DC offset voltage as the stress. Figure 9 presents the time profile *t* of the application voltage to the LC cell for the stress test. The DC voltage applied to the LC cell was fixed at +5 V, and the alternate current (AC) voltage was 3.3 V at 30 Hz, which is the AC voltage showing the 50% transmittance [14–16]. The *I-V* curves obtained by the application of the triangular voltage before and after application of the DC offset voltage are presented in Figure 10. The current maximum and minimum peaks are gradually shifted toward high voltage, indicating that the V_{rDC} is generated under application of the DC offset voltage to the VTN-LC cell and that the V_{rDC} is gradually increased with increasing the period of the stress.

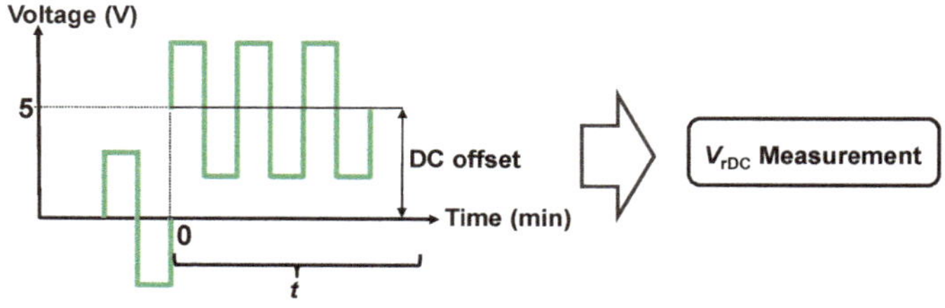

Figure 9. Square waveform of the applied DC offset voltage to the LC cell as the stress test.

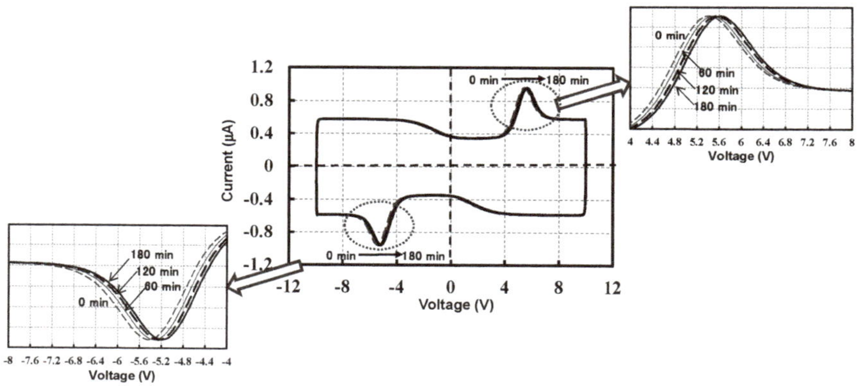

Figure 10. I-V curve shift of the VTN-LC cell with a parameter of time t applying +5 V DC offset voltage as the stress test; the triangular voltage in the range from −10 V to +10 V was applied with 30 Hz at 25 °C.

In Figure 11, the V_{rDC} values determined both from the *I-V* curve shift and flicker minimization are plotted as a function of time *t*. The time dependence of the V_{rDC} determined from the *I-V* curve shift is nearly coincident with the V_{rDC} determined from the flicker minimization. Therefore, the proposed technique, the *I-V* curve shift technique, is useful for the evaluation of the V_{rDC} of the LCDs. Furthermore, we are convinced that the proposed technique would be more convenient than the flicker minimization because the measurements of the *I-V* curve are automatically performed by using the instrumental setup developed by the Toyo Corporation [18].

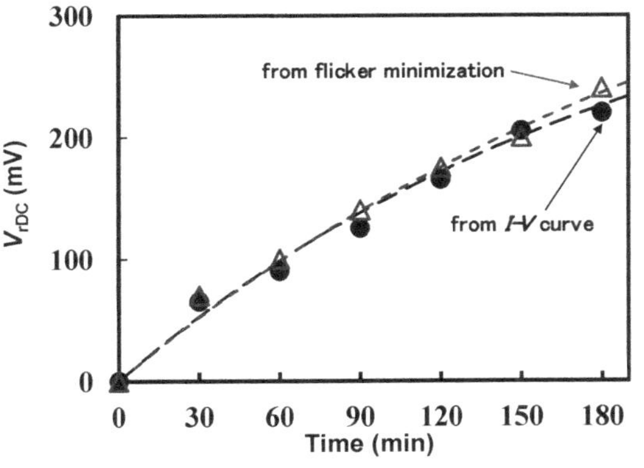

Figure 11. V_{rDC} of the VTN-LC cell as a function of time t applying +5 V DC offset voltage as the stress test; (•) the V_{rDC} determined from the I-V curve shift, and (△) the V_{rDC} determined from the flicker minimization.

5. Measurement of FFS-LC Cell

5.1. I-V Curve

Figure 12 indicates the *I-V* curve of the FFS-LC cell obtained by the application of the 30 Hz triangular voltage at 25 °C. The applied voltage is swept between −10 V to +10 V. The behavior of the FFS-LC cell is different from that of the VTN-LC cell. The *I-V* curve of the FFS-LC cell does not show any specific maximum and/or minimum current peaks. We consider that the current value derived from the rotation of the LC director is so small that no current peak is observed. The rotation of the LC director seems to be slight and limited to the narrow area of the electric field because the electric field concentrates only around the electrodes, as shown in Figure 4. Though we cannot observe maximum and minimum current peaks, a slight shift of the current derived from the anisotropy of the dielectric constant is observed in Figures 12 and 13.

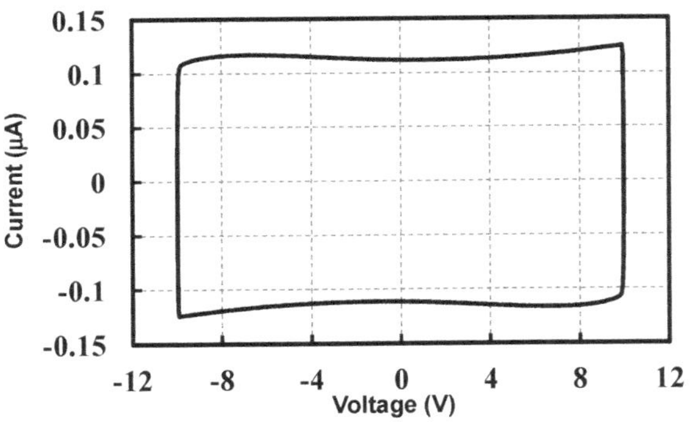

Figure 12. Typical *I-V* curve for the FFS-LC cell; the triangular voltage from −10 V to +10 V was applied with 30 Hz at 25 °C.

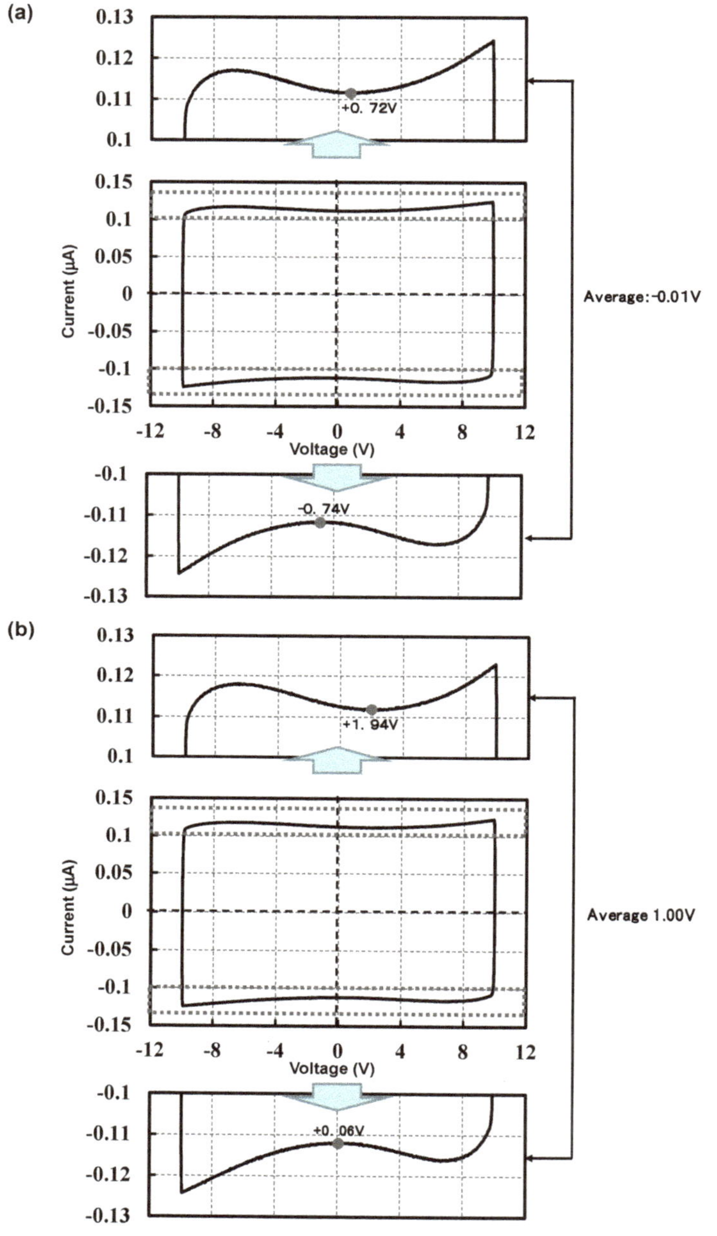

Figure 13. *I-V* curves for the FFS-LC cell (**a**) before and (**b**) after application of the +5 V DC offset voltage for 3 h (180 min) as the stress test; (condition of the measurement) the triangular voltage from −10 V to +10 V was applied with 30 Hz at 25 °C.

5.2. Determination of V_{rDC}

Regarding the FFS-LC cell, the maximum and minimum current peaks did not appear, as described previously. Thus, we considered another method for determining the V_{rDC} from the *I-V* curve. As shown in Figure 13, two slight minimum current peaks were observed for the positive and negative sides. In Figure 13a, two minimum current peaks for the positive and negative sides are +0.72 V and −0.74 V, respectively. The average value of these two voltages is −0.01 V, which is the V_{rDC} before application of the DC offset voltage. Then, the DC offset voltage of +5 V for 180 min was applied to the cell, and the *I-V* curve was measured, as shown in Figure 13b. Two minimum current peaks were also observed, which are +1.94 V and +0.06 V, respectively. Therefore, the average voltage, which indicates the V_{rDC}, becomes 1.00 V. The result indicates that the V_{rDC} of the LC cell with the in-plane electric field can be determined by the method of the *I-V* curve obtained by the application of the triangular voltage.

5.3. V_{rDC} *as a Function of Stress Time Measured by I-V Curve*

The stress test was performed under application of the +5 V DC offset voltage for 180 min, using the FFS-LC cell. The time profile of the stress test is the same as that of Figure 9. The *I-V* curves before and after application of the DC offset voltage are presented in Figure 14. In the case of the FFS-LC cell, the current minimum peaks are gradually shifted toward high voltage, indicating that the V_{rDC} is gradually increased with increasing the stress time. Figure 15 indicates that the V_{rDC} is a function of the stress time, *t*. The result indicates that the V_{rDC} of the FFS-LC cell is gradually increased with increasing the stress time. This is the same tendency with the result of the VTN-LC cell, implying that the V_{rDC} of the FFS mode cell can also be determined by using the *I-V* curve method though it was difficult to determine by the flicker minimization. In the case of the VTN-LC cell, the V_{rDC} determined from the *I-V* curve shift was coincident with the V_{rDC} determined from the flicker minimization technique. Moreover, the V_{rDC} could also be determined for both the VTN- and the FFS-LC cells by the *I-V* curve shift. These results intimate that the V_{rDC} determined from the *I-V* curve shift reflects the internal voltage generated from the adsorption of the ion on the surface of the alignment layer.

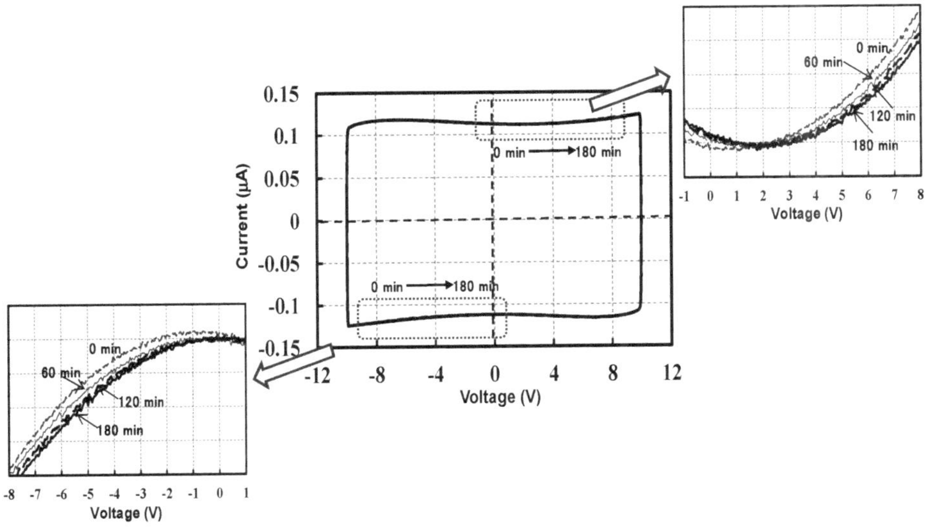

Figure 14. I-V curve shift of the FFS-LC cell with a parameter of time t applying +5 V DC offset voltage as the stress test; the triangular voltage in the range from −10 V to +10 V was applied with 30 Hz at 25 °C.

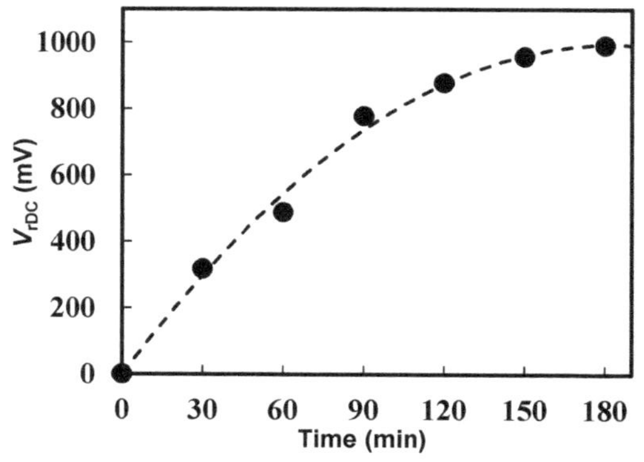

Figure 15. V_{rDC} of the FFS-LC cell as a function of stress time t applying +5 V DC offset voltage.

6. Characteristics of the *I-V* Curve Shift Technique by the Triangular Voltage Application for Determination of the V_{rDC}

In this study, we proposed a novel technique to determine the V_{rDC}. This is the method to evaluate the *I-V* curves by the application of the triangular voltage after application of the DC offset voltage as the stress. The technique is especially useful for the determination of the V_{rDC} because we can not only perform it automatically but also evaluate the V_{rDC} of the FFS mode LC cell, having the lateral electric field. The following table (Table 4) lists the characteristics of the newly proposed technique for the measurement of the V_{rDC}.

Table 4. Proposed evaluation techniques for V_{rDC}.

	I-V Curve Shift
Measuring method	Electrical
Measurement item	Shift of current maximum and/or minimum peaks
Correlation with the image sticking	High
Convenience for measurement	Convenient and easy
Advantage compared with the other techniques	Could be evaluated for both the LC cells with vertically- and in-plane electric fields

7. Conclusions

The V_{rDC} could be determined by the *I-V* curve obtained by the application of the triangular voltage. In the case of the LC cell with the vertical electric field such as the VTN mode, the *I-V* curve shows the maximum and minimum current peaks for the positive and negative sides due to the rotation of the LC director. Therefore, we can determine the V_{rDC} as the average value of these two current peaks. The result of the V_{rDC} evaluated by the proposed method is fairly coincident with the V_{rDC} evaluated by the flicker minimization, which is the optical method. In addition, the proposed method could be applied to the LC cell with the lateral electric field such as the FFS mode though the V_{rDC} could not be determined by the flicker minimization for the FFS mode LC cell. In the case of the FFS mode LC cell, the *I-V* curve shows two minimum current values due to the minimum capacitances of the positive and negative sides. Thus, the V_{rDC} can be determined by using these two minimum current values in a similar manner to the VTN mode LC cell. By using the flicker minimization to the FFS mode LC cell, the V_{rDC} could not be determined because the flicker is generated by not only the generation of V_{rDC} but also the flexoelectric effect. Thus, our proposed technique would be particularly useful for the LC cell with the lateral electric field. Furthermore, the *I-V* curve shift technique has superior characteristics to conventional techniques as shown in Table 4. We believe that the proposed technique is significantly useful for the evaluation of the image sticking of the LCDs.

Author Contributions: Conceptualization, M.M.; Methodology, M.M. and S.I.; Evaluation, M.M.; Simulation, S.I.; Writing-original draft preparation, M.M.; Writing-review and editing, M.M. and S.I. All authors have read and agreed to the published version of the manuscript.

Funding: This research received no external funding.

Institutional Review Board Statement: Not applicable.

Informed Consent Statement: Not applicable.

Data Availability Statement: Not applicable.

Conflicts of Interest: The authors declare no conflict of interest.

References

1. Ishihara, S.; Mizusaki, M. Alignment control technology of liquid crystal molecules. *J. Soc. Inf. Disp.* **2020**, *28*, 44–74. [CrossRef]
2. Schadt, M. Nematic liquid crystals and twisted-nematic LCDs. *Liq. Cryst.* **2015**, *42*, 646–652. [CrossRef]
3. Takeda, A.; Kataoka, S.; Sasaki, T.; Chida, H.; Tsuda, H.; Ohmuro, K.; Koike, Y.; Sasabayashi, T.; Okamoto, K. A super-high image-quality multi-domain vertical alignment LCD by new rubbing-less technology. *SID Int. Symp. Dig. Tech. Pap.* **1998**, *29*, 1077–1079. [CrossRef]
4. Oh-e, M.; Kondo, K. Electro-optical characteristics and switching behavior of the in-plane switching mode. *Appl. Phys. Lett.* **1995**, *67*, 3895–3897. [CrossRef]
5. Lee, S.H.; Lee, S.L.; Kim, H.Y. Electro-optic characteristics and switching principle of a nematic liquid crystal cell controlled by fringe-field switching. *Appl. Phys. Lett.* **1998**, *73*, 2881–2883. [CrossRef]
6. Naemura, S. Liquid-crystal-material technologies for advanced display applications. *J. Soc. Inf. Disp.* **2000**, *8*, 5–9. [CrossRef]
7. Huang, K.T.; Chao, A.; Yu, C.H. Image sticking analysis of different Q-time LC cell by machine vision. *SID Int. Symp. Dig. Tech. Pap.* **2007**, *38*, 665–668. [CrossRef]
8. You, J.; Jung, J.; Choi, N.; Rhie, K.; Shin, S. The study of the correlation between surface anchoring energy and activation energy on patterned vertical alignment (PVA) mode. *SID Int. Symp. Dig. Tech. Pap.* **2007**, *38*, 673–676. [CrossRef]
9. Chen, Y.-C.; Liu, P.-M.; Liao, C.-H.; Ting, T.-L.; Chang, T.-J.; Liao, W.-L.; Lien, A. Releasing behavior of image sticking. *SID Int. Symp. Dig. Tech. Pap.* **2008**, *39*, 1798–1800. [CrossRef]
10. Kamoto, R. Analysis of image sticking on a real MVA cell. *J. Soc. Inf. Disp.* **2008**, *16*, 451–456. [CrossRef]
11. Park, Y.; Kim, S.; Lee, E. A study on reducing image-sticking artifacts in wide-screen TFT-LCD monitors. *J. Soc. Inf. Disp.* **2007**, *15*, 969–973. [CrossRef]
12. Seiberle, H.; Schadt, M. Influence of charge carriers and display parameters on the performance of passively and actively addressed LCDs. *SID Int. Symp. Dig. Tech. Pap.* **1992**, *23*, 25–28.
13. Tanaka, Y.; Goto, Y.; Iimura, Y. Mechanism of DC offset voltage generation in hybrid-aligned nematic liquid crystal displays. *Jpn. J. Appl. Phys.* **1999**, *38 Pt 2*, L1115–L1117. [CrossRef]
14. Mizusaki, M.; Miyashita, T.; Uchida, T.; Yamada, Y.; Ishii, Y.; Mizushima, S. Generation mechanism of residual direct current voltage in a liquid crystal display and its evaluation parameters related to liquid crystal and alignment layer materials. *J. Appl. Phys.* **2007**, *102*, 014904. [CrossRef]
15. Mizusaki, M.; Miyashita, T.; Uchida, T. Behavior of ion affecting image sticking on liquid crystal displays under application of direct current voltage. *J. Appl. Phys.* **2010**, *108*, 104903. [CrossRef]
16. Mizusaki, M.; Miyashita, T.; Uchida, T. Kinetic analysis of image sticking with adsorption and desorption of ions to a surface of an alignment layer. *J. Appl. Phys.* **2012**, *112*, 044510. [CrossRef]
17. Xu, D.; Peng, F.; Chen, H.; Yuan, J.; Wu, S.-T.; Li, M.-C.; Lee, S.-L.; Tsai, W.-C. Image sticking in liquid crystal displays with lateral electric field. *J. Appl. Phys.* **2014**, *116*, 193102. [CrossRef]
18. Inoue, M. Review of various measurement methodologies of migration ion influence on LCD image quality and new measurement proposal beyond LCD materials. *J. Soc. Inf. Disp.* **2020**, *28*, 92–110. [CrossRef]
19. Nishikawa, M.; Suganuma, T.; Tsuda, Y.; Bessho, N.; Iimura, Y.; Kobayashi, S. Properties of voltage holding ratios of liquid crystal cells using organic-solvent-soluble polyimide alignment films. *Jpn. J. Appl. Phys.* **1994**, *33 Pt 2*, L1113–L1116. [CrossRef]
20. Mada, H.; Suzuki, H. Reverse hysteresis loop of nematic liquid crystals in *C-V* characteristic due to static electric field. *Jpn. J. Appl. Phys.* **1987**, *26*, L1092–L1094. [CrossRef]
21. Takahashi, S. The investigation of a dc induced transient optical 30-Hz element in twisted nematic liquid-crystal displays. *J. Appl. Phys.* **1991**, *70*, 5346–5350. [CrossRef]
22. Huang, K.-T.; Huang, Y.-W.; Fang, R.-W.; Chao, Y.-T.; Lee, C.; Lin, S.-C.; Yu, C.-H.; Kao, C.; Jen, T.-S. A turning method to minimizing flicker shift phenomenon in positive FFS mode LCD panel and improve DC image sticking. *SID Int. Symp. Dig. Tech. Pap.* **2018**, *49*, 1711–1714. [CrossRef]
23. Tsuruma, T.; Goto, Y.; Higashi, A.; Watanabe, M.; Yamaguchi, H.; Tomooka, T.; Kikkawa, H. Novel image sticking model in the fringe field switching mode based on the flexoelectric effect. In *31st International Display Research Conference 2011 (EuroDisplay 2011), Arcachon, France, 19–22 September 2011*; Curran Associates, Inc.: Red Hook, NY, USA, 2011; pp. 15–18.

24. Hatsumi, R.; Fukai, S.; Kubota, Y.; Yamashita, A.; Jikumaru, M.; Baba, H.; Moriya, K.; Kubota, D.; Kusunoki, K.; Hirakata, Y.; et al. FFS-mode OS-LCD for reducing eye strain. *J. Soc. Inf. Disp.* **2013**, *21*, 442–450. [CrossRef]
25. Jeong, I.H.; Jang, I.W.; Kim, D.H.; Han, J.S.; Kumar, B.V.; Lee, S.H.; Ahn, S.H.; Cho, S.H.; Yi, C. Investigation on flexoelectric effect in the fringe field switching mode. *SID Int. Symp. Dig. Tech. Pap.* **2013**, *44*, 1368–1371. [CrossRef]
26. Lee, D.-J.; Shim, G.-Y.; Choi, J.-C.; Park, J.-S.; Lee, J.-H.; Baek, J.-H.; Choi, H.C.; Ha, Y.M.; Ranjkesh, A.; Kim, H.-R. Transient flickering behavior in fringe-field switching liquid crystal mode analyzed by positional asymmetric flexoelectric dynamics. *Opt. Express* **2015**, *23*, 34055–34070. [CrossRef] [PubMed]
27. Kim, M.S.; Bos, P.J.; Kim, D.-W.; Yang, D.-K.; Lee, J.H.; Lee, S.H. Flexoelectric effect in an in-plane switching (IPS) liquid crystal cell for low-power consumption display devices. *Sci. Rep.* **2016**, *6*, 35254. [CrossRef]
28. Lee, S.H.; Lee, S.M.; Kim, H.Y.; Kim, J.M.; Hong, S.H.; Jeong, Y.H.; Park, C.H.; Choi, Y.J.; Lee, J.Y.; Koh, J.W.; et al. 18.1″ ultra-FFS TFT-LCD with super image quality and fast response time. *SID Int. Symp. Dig. Tech. Pap.* **2001**, *32*, 484–487. [CrossRef]
29. Miyachi, K.; Kobayashi, K.; Yamada, Y.; Mizushima, S. The world's first photo alignment LCD technology applied to generation ten factory. *SID Int. Symp. Dig. Tech. Pap.* **2010**, *41*, 579–582. [CrossRef]
30. Terashita, S.; Watanabe, K.; Shimoshikiryoh, F. Novel liquid crystal display mode "UV^{2}AII" with photo alignment technology for a large-screen 8K display. *SID Int. Symp. Dig. Tech. Pap.* **2019**, *50*, 62–65. [CrossRef]

MDPI
St. Alban-Anlage 66
4052 Basel
Switzerland
Tel. +41 61 683 77 34
Fax +41 61 302 89 18
www.mdpi.com

Symmetry Editorial Office
E-mail: symmetry@mdpi.com
www.mdpi.com/journal/symmetry

www.ingramcontent.com/pod-product-compliance
Lightning Source LLC
Chambersburg PA
CBHW062051220926
51850CB00010B/2734
9783036583235